METHODS IN MOLECULAR BIOLOGY

Series Editor
John M. Walker
School of Life and Medical Sciences
University of Hertfordshire
Hatfield, Hertfordshire, UK

For further volumes:
http://www.springer.com/series/7651

For over 35 years, biological scientists have come to rely on the research protocols and methodologies in the critically acclaimed Methods in Molecular Biology series. The series was the first to introduce the step-by-step protocols approach that has become the standard in all biomedical protocol publishing. Each protocol is provided in readily-reproducible step-by step fashion, opening with an introductory overview, a list of the materials and reagents needed to complete the experiment, and followed by a detailed procedure that is supported with a helpful notes section offering tips and tricks of the trade as well as troubleshooting advice. These hallmark features were introduced by series editor Dr. John Walker and constitute the key ingredient in each and every volume of the Methods in Molecular Biology series. Tested and trusted, comprehensive and reliable, all protocols from the series are indexed in PubMed.

Manuscript Instructions https://drive.google.com/file/d/1K4Az5vpAFVzYadJ3jKx2-QSo7Dqa_SNA/view?usp=sharing

Mouse Embryogenesis

Methods and Protocols

Second Edition

Edited by

Paul Delgado-Olguín

Translational Medicine, The Hospital for Sick Children, Peter Gilgan Centre for Research and Learning, Toronto, ON, Canada

Humana Press

Editor
Paul Delgado-Olguín
Translational Medicine, The Hospital for Sick Children
Peter Gilgan Centre for Research and Learning
Toronto, ON, Canada

ISSN 1064-3745 ISSN 1940-6029 (electronic)
Methods in Molecular Biology
ISBN 978-1-0716-5125-4 ISBN 978-1-0716-5126-1 (eBook)
https://doi.org/10.1007/978-1-0716-5126-1

This Humana imprint is published by the registered company Springer Science+Business Media, LLC, part of Springer Nature.
The registered company address is: 1 New York Plaza, New York, NY 10004, U.S.A.

If disposing of this product, please recycle the paper.

Preface

The mouse remains our most versatile and powerful model to study mammalian embryogenesis. Its rapid reproductive cycle, amenability to genetic manipulation, and physiological similarity to humans have made it an indispensable system for unraveling the complex mechanisms that orchestrate development. With the ability to conditionally delete, express, or modify genes in defined lineages and at precise developmental time points, we can interrogate gene function with unparalleled spatial and temporal resolution. These capabilities have allowed us not only to elucidate the mechanisms guiding organogenesis and cell differentiation, but also to model congenital disease and understand the molecular basis of postnatal tissue homeostasis.

This volume of the *Methods in Molecular Biology Series* brings together an expanded and updated collection of protocols and review articles designed to empower researchers in their exploration of mouse embryogenesis. Building on the foundation of the first edition, this book now includes new methods and refinements ranging from tissue clearing and three-dimensional imaging, to explant culture systems, lineage tracing, high-resolution transcriptomics, metabolism, and chromatin structure analysis. Each chapter begins with a conceptual overview of the method, followed by a detailed list of materials and step-by-step instructions optimized for reproducibility. Practical notes accompany each protocol, offering insights to navigate technical challenges and improve experimental outcomes.

The chapters are organized according to molecular depth, progressing from foundational methods to approaches that probe cellular and nuclear mechanisms. The protocols presented here share the common goal of revealing how cellular behaviors and molecular pathways are coordinated to shape the embryo. These methods reflect the current frontier of developmental biology and lay the groundwork for future discoveries, as we integrate single-cell approaches, spatial transcriptomics, and an evolutionary perspective into developmental studies.

The scope of this second edition reflects the pace at which our understanding of embryogenesis continues to grow. We hope it serves as a practical resource for both seasoned researchers and newcomers to the field, and that it inspires new questions and experimental strategies to further illuminate the logic of mammalian development.

Toronto, ON, Canada **Paul Delgado-Olguín**

Acknowledgments

This volume reflects the generosity and collaborative spirit of the many investigators who have contributed their trusted methods and expertise for the benefit of the wider research community. Their efforts not only advance the study of mouse embryogenesis but also provide broadly applicable tools for exploring fundamental questions in development, disease, gene regulation, and beyond. I am deeply grateful for their commitment, which has made this volume a truly collective endeavor. My heartfelt thanks to Cristina Mate-Bermejo, whose careful organization and thoughtful editing were indispensable in bringing this work to completion.

Contents

Contributors

AGOSTINO PIERRO • *Translational Medicine, Division of General and Thoracic Surgery, The Hospital for Sick Children, Toronto, ON, Canada*

AMAIRANI CANCINO-BELLO • *Laboratorio de Epigenética de la Regeneración Muscular, Departamento de Genética y Biología Molecular, Centro de Investigación y de Estudios Avanzados del Instituto Politécnico Nacional, Ciudad de México, México*

ANDREW L. LOPEZ III • *Department of Biomedical Engineering, University of Houston, Houston, TX, USA; Department of Integrative Physiology, Baylor College of Medicine, Houston, TX, USA*

AZADEH YEGANEH • *Translational Medicine, The Hospital for Sick Children, Peter Gilgan Centre for Research and Learning, Toronto, ON, Canada*

BEHZAD YEGANEH • *Apoptosis Research Centre, Children's Hospital of Eastern Ontario Research Institute, Ottawa, ON, Canada; Department of Cellular and Molecular Medicine, University of Ottawa, Ottawa, ON, Canada*

DOROTHY LEE • *Translational Medicine, Division of General and Thoracic Surgery, The Hospital for Sick Children, Toronto, ON, Canada*

EDUARDO BLANCO-OLAIS • *Molecular Genetics Department, Instituto de Fisiología Celular, Universidad Nacional Autónoma de México, México City, Mexico*

ELENA DE LA CRUZ HERRERA-COGCO • *Laboratorio de Farmacología Cardiovascular, Facultad de Ciencias Químicas, Universidad Veracruzana, Orizaba, Veracruz, México*

ENRIQUE MÉNDEZ-BOLAINA • *Facultad de Ciencias Químicas, Universidad Veracruzana, Orizaba, Veracruz, México*

ESMERALDA LIRA-ROMERO • *Laboratorio de Biología Molecular, Escuela de Medicina, Universidad Panamericana, México City, México; Sección de Estudios de Posgrado e Investigación, Escuela Superior de Medicina, Instituto Politécnico Nacional, México City, México*

ESTELA G. GARCÍA-GONZÁLEZ • *Laboratorio de Epigenética de la Regeneración Muscular, Departamento de Genética y Biología Molecular, Centro de Investigación y de Estudios Avanzados del Instituto Politécnico Nacional, Ciudad de México, México*

FRANCISCO JAVIER ESTRADA-MENA • *Laboratorio de Biología Molecular, Escuela de Medicina, Universidad Panamericana, México City, México*

HAMNA AMMAR • *Translational Medicine, The Hospital for Sick Children, Peter Gilgan Centre for Research and Learning, Toronto, ON, Canada*

HOON-KI SUNG • *Department of Laboratory Medicine and Pathobiology, University of Toronto, Toronto, ON, Canada; Translational Medicine, The Hospital for Sick Children, Peter Gilgan Centre for Research and Learning, Toronto, ON, Canada*

IRINA V. LARINA • *Department of Integrative Physiology, Baylor College of Medicine, Houston, TX, USA*

ISRAEL RAMÍREZ-SÁNCHEZ • *Sección de Estudios de Posgrado e Investigación, Escuela Superior de Medicina, Instituto Politécnico Nacional, México City, México*

J. MANUEL HERNÁNDEZ-HERNÁNDEZ • *Laboratorio de Epigenética de la Regeneración Muscular, Departamento de Genética y Biología Molecular, Centro de Investigación y de Estudios Avanzados del Instituto Politécnico Nacional, Ciudad de México, México*

JENNIFER A. MITCHELL • *Department of Cell and Systems Biology, University of Toronto, Toronto, ON, Canada; Department of Laboratory Medicine and Pathobiology, University of Toronto, Toronto, ON, Canada*
JIELIN YANG • *Translational Medicine, The Hospital for Sick Children, Peter Gilgan Centre for Research and Learning, Toronto, ON, Canada*
JU HEE LEE • *Department of Laboratory Medicine and Pathobiology, University of Toronto, Toronto, ON, Canada; Translational Medicine, The Hospital for Sick Children, Peter Gilgan Centre for Research and Learning, Toronto, ON, Canada*
JULIE PAN • *Heart Institute, Department of Cellular and Molecular Medicine, Faculty of Medicine, University of Ottawa, Ottawa, ON, Canada*
KAFI EALEY • *School of Nutrition, Toronto Metropolitan University, Toronto, ON, Canada*
KARINA JÁCOME-LÓPEZ • *Molecular Genetics Department, Instituto de Fisiología Celular, Universidad Nacional Autónoma de México, México City, Mexico*
KAZUKO KOSHIBA-TAKEUCHI • *Faculty of Life Sciences, Toyo University, Asaka, Saitama, Japan*
KYOUNG-HAN KIM • *Heart Institute, Department of Cellular and Molecular Medicine, Faculty of Medicine, University of Ottawa, Ottawa, ON, Canada*
LIJUN CHI • *Translational Medicine, The Hospital for Sick Children, Peter Gilgan Centre for Research and Learning, Toronto, ON, Canada*
MAGALLY RAMÍREZ-RAMÍREZ • *Laboratorio de Epigenética de la Regeneración Muscular, Departamento de Genética y Biología Molecular, Centro de Investigación y de Estudios Avanzados del Instituto Politécnico Nacional, Ciudad de México, México*
MARÍA LUISA BENÍTEZ-HESS • *Laboratorio de Epigenética de la Regeneración Muscular, Departamento de Genética y Biología Molecular, Centro de Investigación y de Estudios Avanzados del Instituto Politécnico Nacional, Ciudad de México, México*
MARIIA CHEREDNYCHENKO • *Department of Cell and Systems Biology, University of Toronto, Toronto, ON, Canada*
MARIO FILICE • *Department of Cell and Systems Biology, University of Toronto, Toronto, ON, Canada*
MARTIN POST • *Department of Laboratory Medicine and Pathobiology, University of Toronto, Toronto, ON, Canada; Translational Medicine, The Hospital for Sick Children, Peter Gilgan Centre for Research and Learning, Toronto, ON, Canada*
MAURICIO HERNÁNDEZ-SOMILLEDA • *Laboratorio de Epigenética de la Regeneración Muscular, Departamento de Genética y Biología Molecular, Centro de Investigación y de Estudios Avanzados del Instituto Politécnico Nacional, Ciudad de México, México*
MAYRA FURLAN-MAGARIL • *Molecular Genetics Department, Instituto de Fisiología Celular, Universidad Nacional Autónoma de México, México City, Mexico*
MIRANDA PANNELLA • *Translational Medicine, The Hospital for Sick Children, Peter Gilgan Centre for Research and Learning, Toronto, ON, Canada*
NAHO FUJIWARA • *Department of Pediatric General and Urogenital Surgery, Juntendo University Graduate School of Medicine, Tokyo, Japan; Translational Medicine, Division of General and Thoracic Surgery, The Hospital for Sick Children, Toronto, ON, Canada*
NATHANIEL VO • *Department of Laboratory Medicine and Pathobiology, University of Toronto, Toronto, ON, Canada; Translational Medicine, The Hospital for Sick Children, Peter Gilgan Centre for Research and Learning, Toronto, ON, Canada*

NAVROOP K. DHALIWAL • *Department of Cell and Systems Biology, University of Toronto, Toronto, ON, Canada; Developmental and Stem Cell Biology, The Hospital for Sick Children, Peter Gilgan Centre for Research and Learning, Toronto, ON, Canada*
PAUL DELGADO-OLGUÍN • *Department of Molecular Genetics, University of Toronto, Toronto, ON, Canada; Heart & Stroke Richard Lewar Centre of Excellence in Cardiovascular Research, Toronto, ON, Canada; Translational Medicine, The Hospital for Sick Children, Peter Gilgan Centre for Research and Learning, Toronto, ON, Canada*
QIWEI ZHANG • *Department of Laboratory Medicine and Pathobiology, University of Toronto, Toronto, ON, Canada; Translational Medicine, The Hospital for Sick Children, Peter Gilgan Centre for Research and Learning, Toronto, ON, Canada*
QUETZALCOATL ESCALANTE-COVARRUBIAS • *Department of Molecular Genetics, University of Toronto, Toronto, ON, Canada; Translational Medicine, The Hospital for Sick Children, Peter Gilgan Centre for Research and Learning, Toronto, ON, Canada*
RIMSHAH ABID • *Heart Institute, Department of Cellular and Molecular Medicine, Faculty of Medicine, University of Ottawa, Ottawa, ON, Canada*
ROSARIO PÉREZ-MOLINA • *Molecular Genetics Department, Instituto de Fisiología Celular, Universidad Nacional Autónoma de México, México City, Mexico*
SALONI MODI • *Department of Cell and Systems Biology, University of Toronto, Toronto, ON, Canada*
SANDRA VUONG • *Translational Medicine, The Hospital for Sick Children, Peter Gilgan Centre for Research and Learning, Toronto, ON, Canada*
WATARU KATANO • *Division of Molecular Cardiovascular Biology, Cincinnati Children's Hospital Medical Center, Cincinnati, OH, USA; Faculty of Life Sciences, Toyo University, Asaka, Saitama, Japan*
YENA OH • *Heart Institute, Department of Cellular and Molecular Medicine, Faculty of Medicine, University of Ottawa, Ottawa, ON, Canada*
YUKI TAJIKA • *Department of Radiological Technology, Gunma Prefectural College of Health Sciences, Maebashi, Gunma, Japan*
YUNHAN LI • *Translational Medicine, The Hospital for Sick Children, Peter Gilgan Centre for Research and Learning, Toronto, ON, Canada*
ZOE E. GILLESPIE • *Department of Cell and Systems Biology, University of Toronto, Toronto, ON, Canada*

Check for updates

Chapter 1

Mouse Genotyping

Yunhan Li, Sandra Vuong, Lijun Chi and Paul Delgado-Olguín

Abstract

Genotyping is a valuable tool for identifying organisms that carry genetic variations, including insertions and deletions. This method involves extracting DNA from animal tissue samples and subsequently amplifying genomic regions of interest using the polymerase chain reaction (PCR). The amplified product is analyzed by agarose gel electrophoresis based on DNA size.

Keywords Genotype, DNA, Polymerase chain reaction

1 Introduction

There are about 24,000 protein-coding genes in the mouse genome, and the functions of many of them have yet to be elucidated [1]. In a realm where genetically modified mouse lines are instrumental to discovering the function of unknown genes, identifying mice carrying genetic modifications affecting the function of specific genes is required.

Genotyping mice involves the amplification of a genomic region of interest from isolated DNA. The DNA is commonly extracted from a small piece of tissue, most commonly from a tail clip, ear notch, or yolk sac. Primers designed to target the genomic region or gene of interest are used for amplification by the polymerase chain reaction (PCR) [2–4]. Primers can be designed using many freely available software platforms such as PrimerBLAST [5, 6]. The PCR amplified products are then separated by gel electrophoresis and visualized using a gel documentation system. Genotypes are then determined by identifying PCR products of the corresponding expected size(s) based on the locations of the primers used.

This chapter describes a rapid and cost-effective method for DNA extraction that bypasses protein digestion steps. Although this quick method may require optimization with certain primer pairs, it consistently yields clean and specific PCR products. Based on

Paul Delgado-Olguín (ed.), *Mouse Embryogenesis*, Methods in Molecular Biology, vol. 3010,
https://doi.org/10.1007/978-1-0716-5126-1_1,

our experience, this technique is effective for DNA extraction from both embryonic yolk sacs and postnatal ear notch samples, making it suitable for a broad range of genotyping applications in mouse models [7–12].

2 Materials

All solutions should be prepared using double-distilled water (MilliQ water). Solutions can be prepared in advance and stored at room temperature unless otherwise stated.

2.1 Quick DNA Isolation

1. *50 mM NaOH:* Add 250 μL of 10 N sodium hydroxide to 49.75 mL of water for a total of 50 mL.
2. *0.5 M Tris–HCl pH 8.0:* To prepare 1 M Tris–HCl pH 8.0 stock solution, weigh 60.55 g of Tris base and dissolve in 400 mL of water. Adjust the pH to 8.0 by adding 37% concentrated Hydrochloric acid to the solution. Allow the solution to acclimate to room temperature before bringing the volume of the solution to 500 mL with water. Add 25 mL of water to 25 mL of 1 M Tris–HCl solution to make 50 mL of 0.5 M Tris–HCl pH 8.0.

2.2 Polymerase Chain Reaction (PCR)

1. *10× PCR Buffer:* 100 mM Tris pH 8.3, 500 mM KCl, 20 mM $MgCl_2$. To prepare a 1 M Tris pH 8.3 stock solution, dissolve 60.55 g of Tris base in 400 mL of water. Adjust pH to 8.3 by slowly adding concentrated HCl, 37% in solution. Bring the solution to room temperature before adjusting the final volume to 500 mL with water. To make a 1 M KCl stock solution, dissolve 37.275 g of potassium chloride in 400 mL of water. Bring the final volume up to 500 mL and autoclave for 20 min on a liquid cycle. To prepare 1 M $MgCl_2$ stock solution, weigh 101.65 g of magnesium chloride hexahydrate and dissolve in 400 mL of water. Bring the final volume to 500 mL. Autoclave for 20 min on liquid cycle.
   To make a 20 mL solution of 10× PCR Buffer, add 2 mL of 1 M Tris pH 8.3, 10 mL 1 M KCl, and 2.4 mL 1 M $MgCl_2$ to 5.6 mL of water. Aliquot 1 mL of 10× PCR buffer into 20 1.5 mL microcentrifuge tubes for storage. Store at −20°C (*see* **Note 1**).
2. *2 mM dNTPs*: Combine 250 μL of 100 mM dATP, dCTP, dGTP, and dTTP in 11.5 mL of water for a total volume of 12.5 mL. Aliquot 250 μL of 2 mM dNTPs into 50 1.5 mL microcentrifuge tubes for storage. Store at −20°C (*see* **Note 2**).
3. *5 μM Primer Pair Solution*: First, make 100 μM primer stock solutions of both forward and reverse oligonucleotides. To

make 100 μL of 5 μM working primer pair solution, add 10 μL of forward primer and 10 μL of reverse primer to 80 μL of water. Store at −20°C.

4. *Taq Polymerase*: Taq polymerase may be commercially purchased (*see* **Note 3**).
5. 6× Orange G Loading dye: 0.02% Orange G (weight/volume) in 30% sucrose. Sucrose solution can be prepared by dissolving 15 g of sucrose in water. To make loading dye, weigh 2 mg of Orange G dye and dissolve it in 10 mL of 30% sucrose (*see* **Note 4**).

2.3 Gel Electrophoresis

1. *TAE Buffer:* Prepare 1 L of 50× TAE Buffer stock solution by dissolving 24.2 g of Tris Base in 700 mL of water. Add 100 mL of 0.5 M EDTA, pH 8.0, and 57.1 mL of Glacial Acetic Acid to the solution. Bring the solution to a final volume of 1 L. To make 1× TAE Buffer, add 20 mL of 50× TAE solution to 980 mL of water.
2. *Agarose* (*see* **Note 5**).
3. *Nucleic Acid Gel Staining Solution* (*see* **Note 6**).
4. *DNA Ladder* (*see* **Note 7**).

3 Methods

All steps should be carried out at room temperature unless otherwise stated. The quick method is recommended for both embryonic tissues and DNA isolation from ear notches and tail clips.

3.1 Sample Collection

1. For postnatal or adult mice, a single ear notch is sufficient for DNA extraction. For embryos, either an embryonic yolk sac or approximately 3 mm of tail tissue can be used. After collecting the sample, place it into a sterile 1.5 mL microcentrifuge tube (*see* **Note 8**).

3.2 DNA Extraction

Quick DNA Extraction

1. To each microcentrifuge tube, add 150 μL of 50 mM NaOH. Make sure that the sample is submerged in the solution (*see* **Note 9**).
2. Incubate samples at 95°C for 45 min, vortex at the 30-min mark to disaggregate tissue (*see* **Note 10**).
3. Add 60 μL of 0.5 M Tris–HCl pH 8.0 to neutralize pH and stop the reaction, corresponding to 40% of the NaOH volume used. Mix well. The sample is now ready for PCR (*see* **Note 11**).

3.3 Polymerase Chain Reaction

1. Thaw 10× PCR buffer, 2 mM dNTPs, and primer working solution(s), then keep them on ice
2. In sterile PCR tube(s), set up each reaction (total volume 20 μL) as follows:
   2 μL 10× PCR Buffer, 2 μL 2 mM dNTPs, 2 μL primer mix, 2 μL *Taq* Polymerase, 8 μL water, and 4 μL DNA sample (*see* **Note 12**).
3. Set up a PCR program on a thermocycler as follows:
   (a) 5:00 min at 94°C. This step is for the initial denaturation of DNA.
   (b) 0:30 s at 94°C. This step is for the denaturation of DNA.
   (c) 0:30 s at a specific annealing temperature for primers (*see* **Note 13**).
   (d) 1:00 min at 72°C. This step allows for elongation of DNA strands (*see* **Note 14**).
   (e) Repeat Steps 2–4 for a total of 35 cycles. This step allows for exponential amplification of the DNA template (*see* **Note 15**).
   (f) 5:00 min at 72°C. Final extension of DNA templates occurs here.
   (g) 10:00 min at 4°C. PCR products may then be removed and stored at 4°C until analysis (*see* **Note 16**).

3.4 Gel Electrophoresis

1. Weigh enough agarose to prepare the required volume and percentage. The percentage of the gel should be chosen based on the DNA fragment size. To separate small DNA fragments, a higher percentage is required. Measure the required amount of TAE buffer to cast an agarose gel. Mix the agarose and TAE buffer in a clean flask (*see* **Note 17**).
2. Microwave to heat the solution to fully dissolve the agarose in TAE buffer (*see* **Note 18**).
3. Let the solution cool down at room temperature. After a few minutes, add the nucleic acid gel stain to the mixture. Mix well. When the flask is cool enough to be touched by a gloved hand, it is ready for pouring into a gel cast (*see* **Note 19**).
4. Let gel polymerize at room temperature for a minimum of 20 min (*see* **Note 20**).
5. Remove the comb from the gel and place the gel into an electrophoresis tank. Fill the tank with enough TAE buffer to cover the gel. Load samples and the DNA ladder into the wells (*see* **Note 21**).
6. Let the samples run at 100 V for 45 min (*see* **Note 22**).
7. Image the gel using a gel documentation system (*see* **Note 23**).

4 Notes

1. 10× PCR reaction buffer may be purchased with *Taq Polymerase.*
2. 2 mM dNTPs may be replaced with the dNTP mix included with commercially purchased *Taq Polymerase.* Remember to take note of the concentration of the commercial dNTP reagent mix and make adjustments to the PCR recipe accordingly.
3. As an alternative to purchasing commercial *Taq* polymerase, the enzyme can be purified in the lab using a published protocol [13].
4. Orange G dye may be replaced with other dyes, such as bromophenol blue and xylene cyanol.
5. Molecular Biology Grade Agarose used from FroggaBio.
6. Commercially available nucleic acid gel staining solutions that have been designed as safer alternatives to ethidium bromide (EtBr), such as *RedSafe™*, *GelRed*, and *SYBR Safe*, are recommended.
7. DNA ladders can be commercially purchased (FroggaBio). The size of the DNA ladder used depends on the size of the PCR product. For example, 100 bp DNA ladders are recommended for PCR templates that are shorter than 1500 bp. 1 kb DNA ladders should be used for templates that are a few kilobases long.
8. As an alternative to a tail sample, the hindlimb or forelimb from embryos of embryonic day (E) 10.5 or older can be used. Yolk sac membranes may also be used. For pups that are younger than 21 days (P21), either a tail sample of approximately 5 mm or a single ear notch is sufficient. For adult mice, 2 mm ear notch samples should be collected. While fresh samples are ideal for extraction, samples may also be stored at −20°C for future DNA extraction.
9. The amount of 50 mM NaOH added is adjusted based on the sample size. For samples that are smaller than approximately 2 mm, 90 μL of 50 mM NaOH is added. For samples larger than 7 mm, the volume of 50 mM NaOH is increased to 300 μL. For individual 2 mm ear notches, 150 μL 50 mM NaOH is added.
10. Incubation time can be extended based on the samples used. An incubation period of 10–15 min is ideal for fresh embryonic tissue. For ear notches, incubation of 30–45 min is best. The degree of tissue disaggregation can be periodically checked, and the tubes can be manually agitated for better

digestion. Samples may be removed from 95°C when all the tissue has been disintegrated.

11. To quench the reaction, the amount of 0.5 M Tris–HCl, pH 8.0, required is 40% the volume of NaOH used. Samples can be stored at 4°C for short term or −20°C for long-term.

12. If analyzing more than one sample, preparing a master mix with all components excluding DNA is recommended. Additionally, 4 μL of 1× loading dye can be added to each reaction at this step to increase the total reaction volume to 24 μL. Water and *Taq* polymerase volumes can be adjusted according to the manufacturer's instructions.

13. To determine the optimal annealing temperature for a specific primer pair, running an initial gradient PCR is recommended. Usually, a gradient PCR with temperatures ranging from 53 to 70°C is sufficient to determine the optimal primer annealing temperature. For PCR in Fig. 1, we used forward primers (F) for wild type allele: ATGATAAAGAATAAAGTCCCCGAGTAG, F for floxed allele: GGAACCCTTCGAAGTTCCTATT, and reverse primer (R): GCAACTCCACCTGAATCTCTT, which amplify a 300 bp wild type band, and a floxed band of 395 bp of the gene encoding the *Tbx15* [8].

14. An extension time of 1:00 min is sufficient for amplification of products expected to 1 kb in length. For longer fragments, extension time may be increased. As a general rule for calculation, use an extension time of 0:30 s for every 500 bp to be amplified.

15. The total number of cycles may be adjusted to increase or decrease final template concentration. 25–35 cycles are usually sufficient for amplification. Increasing the reaction over 35 cycles is not recommended, as the likelihood of obtaining unspecific bands increases.

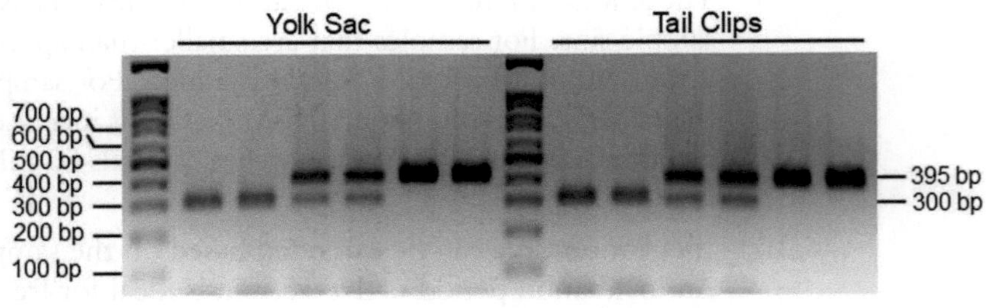

Fig. 1 Genotyping of samples using the quick method with the same primer pair. The DNA isolated from the yolk sacs of E10.5 (left) and the tail clips of postnatal 2 weeks (right). PCR amplification of the Tbx15 gene was performed, with the floxed allele detected at 395 bp and the wild-type allele at 300 bp. A 3 μL of 100 bp DNA ladder was used as a size reference. The 2% gel was electrophoresed at 100 V for 90 min

16. PCR products may be stored at 4°C for a period of up to 1 week before gel electrophoresis. Alternatively, PCR products may be stored at −20°C for long term.
17. The percentage of agarose gel should be chosen based on the size of the DNA fragment that needs to be resolved. For example, 1.5% (weight/volume) agarose gels resolve fragments between 200 bp and 3000 bp. Keep in mind that the agarose percentage adequate for DNA size range may differ depending on agarose formulation. The total amount of agarose and 1× TAE buffer needed will vary based on the size of the gel and the electrophoresis equipment.
18. For higher percentage gels (>2%), it is better to heat the agarose at a lower heating power to prevent TAE evaporation. Higher percentage gels will also take longer to dissolve fully compared to lower percentage gels. It is recommended that mixtures be heated for an initial minute and then swirled for a few seconds before additional heating. The length of heating will vary based on the microwave used and should be optimized accordingly.
19. When handling higher percentage gels, it is best to add the appropriate amount of nucleic acid gel stain no longer than 5 min after the flask has been removed from the microwave. This is to ensure the even incorporation of nucleic acid stain into the gel.
20. The period needed for the agarose gel to fully polymerize will vary depending on the gel's percentage. Higher percentage gels will polymerize faster, and lower percentage gels will take longer to solidify. Ensure gels are completely polymerized before removing the comb and gel cast. Precast gels may be stored overnight at 4°C for running the next day. Make sure to let gels acclimatize to room temperature before use after retrieval from 4°C storage.
21. If DNA loading dye was not added to the PCR master mix, add DNA loading dye to the samples before loading them into the wells.
22. Voltage and running length may be adjusted based on equipment and band resolution needed. For example, separation of bands that are 20 base pairs apart in length requires running the full gel length, whereas separation of bands a few hundred base pairs apart can be run for half the length of a gel.
23. Gel documentation system from BIO-RAD using Image Lab Software.

Acknowledgments

This work was supported by the Canadian Institutes of Health Research (CIHR) (162208, 149046, and 468633) and the Heart & Stroke Foundation of Canada (G-25-0040100) to P.D.-O.

Conflict of Interest The authors have no conflicts of interest to declare that are relevant to the content of this chapter.

References

1. Mouse Genome Sequencing Consortium, Waterston RH, Lindblad-Toh K, Birney E, Rogers J, Abril JF et al (2002) Initial sequencing and comparative analysis of the mouse genome. Nature 420(6915):520–562
2. Newton CR, Graham A, Heptinstall LE, Powell SJ, Summers C, Kalsheker N, Smith JC, Markham AF (1989) Analysis of any point mutation in DNA. The amplification refractory mutation system (ARMS). Nucleic Acids Res 17(7):2503–2516
3. Saiki RK, Gelfand DH, Stoffel S, Scharf SJ, Higuchi R, Horn GT, Mullis KB, Erlich HA (1988) Primer-directed enzymatic amplification of DNA with a thermostable DNA polymerase. Science 239(4839):487–491
4. Saiki RK, Scharf S, Faloona F, Mullis KB, Horn GT, Erlich HA, Arnheim N (1985) Enzymatic amplification of beta-globin genomic sequences and restriction site analysis for diagnosis of sickle cell anemia. Science 230(4732):1350–1354
5. Kim TD (2000) PCR primer design: an inquiry-based introduction to bioinformatics on the World Wide Web. Biochem Mol Biol Educ 28(5):274–276
6. Ye J, Coulouris G, Zaretskaya I, Cutcutache I, Rozen S, Madden TL (2012) Primer-BLAST: a tool to design target-specific primers for polymerase chain reaction. BMC Bioinf 13:134
7. Ahmed A, Delgado-Olguin P (2020) Embryonic programming of heart disease in response to obesity during pregnancy. Biochim Biophys Acta Mol Basis Dis 1866(2):165402
8. Ahmed A, Syed JN, Chi L, Wang Y, Perez-Romero C, Lee D, Kocaqi E, Caballero A, Yang J, Escalante-Covarrubias Q, Ishimura A, Suzuki T, Aguilar-Arnal L, Gonzales GB, Kim KH, Delgado-Olguín P (2023) KDM8 epigenetically controls cardiac metabolism to prevent initiation of dilated cardiomyopathy. Nat Cardiovasc Res 2(2):174–191
9. Chi L, Ahmed A, Roy AR, Vuong S, Cahill LS, Caporiccio L, Sled JG, Caniggia I, Wilson MD, Delgado-Olguin P (2017) G9a controls placental vascular maturation by activating the Notch Pathway. Development 144(11):1976–1987
10. Chi L, Zhong L, Lee D, Yu X, Caballero A, Nieman B, Delgado-Olguín P (2023) G9a inactivation in progenitor cells with Isl1-Cre with reduced recombinase activity models aspects of Dandy-Walker complex. Biol Open 12(8):bio059894
11. Delgado-Olguín P, Huang Y, Li X, Christodoulou D, Seidman CE, Seidman JG, Tarakhovsky A, Bruneau BG (2012) Epigenetic repression of cardiac progenitor gene expression by Ezh2 is required for postnatal cardiac homeostasis. Nat Genet 44(3):343–347
12. Roy AR, Ahmed A, DiStefano PV, Chi L, Khyzha N, Galjart N, Wilson MD, Fish JE, Delgado-Olguín P (2018) The transcriptional regulator CCCTC-binding factor limits oxidative stress in endothelial cells. J Biol Chem 293(22):8449–8461
13. Pluthero FG (1993) Rapid purification of high-activity Taq DNA polymerase. Nucleic Acids Res 21(20):4850–4851

Check for updates

Chapter 2

Dynamic Imaging of Mouse Embryos and Cardiodynamics in Static Culture

Andrew L. Lopez III and Irina V. Larina

Abstract

The heart is a dynamic organ that undergoes rapid morphological and mechanical changes during early embryonic development. Characterizing these early moments is important for our understanding of proper embryonic development and the treatment of heart disease. Traditionally, tomographic imaging modalities and fluorescence-based microscopy are excellent approaches for visualizing structural features and gene expression patterns, respectively, and connecting aberrant gene programs to pathological phenotypes. However, these approaches typically require static samples or fluorescent markers, which can limit the amount of information we can derive from the dynamic and mechanical changes that regulate heart development. Optical coherence tomography (OCT) is unique in this context because it enables the acquisition of three-dimensional structural images and four-dimensional (3D + time) functional imaging of living mouse embryos without the need for fixation or contrast agents. In this chapter, we focus on how OCT can visualize heart morphology at different stages of development and provide cardiodynamic information to reveal the mechanical properties of the developing heart.

Keywords Optical coherence tomography, Cardiovascular development, Embryo culture, Heart morphogenesis, Cardiodynamic analysis, Mouse, Live imaging

1 Introduction

The vertebrate embryonic heart is the first functional organ in development, and is vital to support the growing nutritional demands of the developing embryo. To sustain development, the early heart must execute quick morphological and mechanical changes to achieve proper configuration, increase cardiac output, and maintain oxygenation homeostasis. Failure to achieve these changes can have many developmental implications, ranging from embryonic death to congenital heart disease, which can mildly or severely impact post-uterine development and the ability to thrive. Investigating the regulatory mechanisms involved in this process is essential to understand correct cardiovascular development, Cardiovascular development, and generate therapies to treat heart abnormalities.

Paul Delgado-Olguín (ed.), *Mouse Embryogenesis*, Methods in Molecular Biology, vol. 3010,
https://doi.org/10.1007/978-1-0716-5126-1_2,

Our current understanding of embryonic cardiovascular development relies heavily on molecular analysis, histology, and tomographic and fluorescence-based imaging. These approaches have revealed valuable information in regard to gene programs and connecting them to normal developmental events or assigning them to aberrant phenotypes. While powerful, these approaches lack functional in vivo analysis to understand how biomechanics influence proper cardiogenesis. To this end, live embryo culture methods, coupled with the use of optical coherence tomography (OCT) and advanced post-processing algorithms, can reveal information about the mechanical environment in the embryonic heart [1–4].

In this chapter, we use the mouse as a model for mammalian embryonic cardiovascular development. We will describe live embryo culture, methods for OCT imaging to analyze live embryos, assess cardiodynamics, and quantify hemodynamics in the developing heart.

2 Materials

2.1 Embryo-Dissection Medium

1. DMEM/F12: 2.5 mM L-Glutamine and 15 mM HEPES buffer.
2. Fetal bovine serum (FBS), stored in 5 mL aliquots at −30°C.
3. Penicillin/streptomycin, 100×, stored as 5 mL aliquots at −30°C.

2.2 Rat Serum Extraction

1. Sprague Dawley adult male rats.
2. Ether.
3. Vacutainer blood collection tubes.
4. Vacutainer blood collection sets.
5. 0.45 μm syringe filter.

2.3 Embryo-Dissection

1. Two #5 micro dissecting forceps.
2. Micro dissecting scissors.
3. Micro dissecting tweezers.
4. 35 mm petri dishes.

2.4 Embryo Culture Medium

1. DMEM/F12 with 2.5 mM L-Glutamine and 15 mM HEPES buffer.
2. Rat serum (stored as 1 mL aliquots at −80°C).
3. Penicillin/streptomycin, 100× stored as 5 mL aliquots at −30°C.

3 Methods

3.1 Rat-Serum Extraction (see Note 1)

1. Anesthetize the rat with ether (*see* **Note 2**).
2. Expose the dorsal aorta by abdominal incision with scissors.
3. Collect blood into a Vacutainer blood collection tube using the Vacutainer blood collection set.
4. Once blood is collected, invert the tube a few times to mix and keep it on ice while extracting blood from other rats.
5. Centrifuge tubes with collected blood at 1300 × *g* for 20 min. The supernatant should look golden. If the supernatant looks pink, which indicates erythrocyte lysis, we recommend discarding those tubes. Pool supernatant (serum) into 15 mL centrifuge tubes.
6. Centrifuge serum at 1300 × *g* for 10 min to pellet down the remaining erythrocytes. Pool serum into 50 mL centrifuge tubes.
7. Heat-inactivate serum at 56°C for 30 min in a water bath. Keep the tube lid unscrewed for ether evaporation.
8. Keep serum overnight at 4°C with the lid unscrewed for further evaporation.
9. Filter serum through a 0.45 μm syringe filter.
10. Aliquot serum by 1 mL and keep at −80°C.

3.2 Embryo Dissections

1. Set up matings and check for vaginal plugs every morning. If the plug is found, the pregnancy is taken as E0.5.
2. On the appropriate embryonic day, prepare the embryo dissection medium fresh. Approximately 100 mL of medium will be used per litter.
3. Preheat dissection medium to 37°C.
4. Preheat the dissection station and maintain it at 37°C. We use a custom-built heating station (Fig. 1a) composed of a plexiglass chamber surrounding a dissection microscope, a commercial space heater, and a temperature controller.
5. Sacrifice the pregnant mouse in accordance with established guidelines.
6. Dissect the uterus with embryos and transfer over to warm (37°C) dissection medium in a petri dish. Immediately move the dish into the temperature-controlled dissection station.
7. With micro dissecting forceps, under microscope guidance, sequentially separate individual embryos from the muscular uterine tissue.

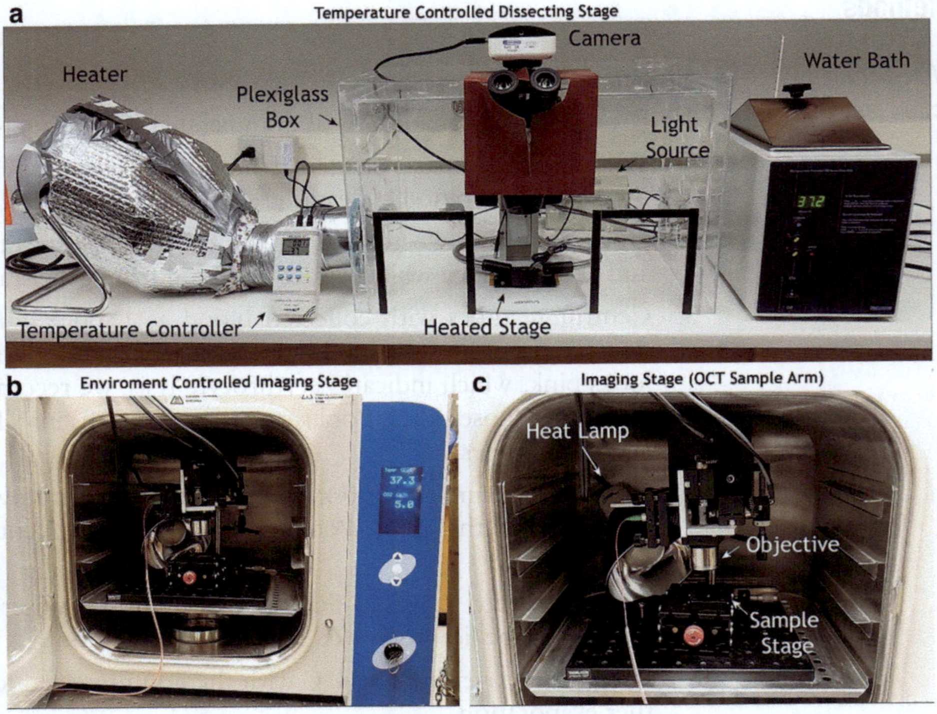

Fig. 1 Dissection and imaging stage. (**a**) Embryos are dissected in a heated lab-built dissection stage consisting of a plexiglass box, an Arduino-controlled heated stage, and a commercial heater connected to the plexiglass box and modulated by a temperature controller. (**b**) Dissected embryos are imaged in a commercial incubator maintaining 37°C and 5%CO_2. (**c**) The OCT sample arm resides within the incubator to image live embryos. Heating lamps are used to maintain heat while the incubator is open

8. Delicately peel away the deciduum covering the embryo. Leave a large piece of the deciduum connected to the ectoplacental cone. This remaining tissue can be shaped with dissection forceps to position the embryo with the heart toward the imaging beam.
9. Delicately peel away Reichert's membrane, exposing the yolk sac, while leaving the yolk sac intact (*see* **Note 3**).
10. Transfer the embryo to a culture dish with fresh warm media using a transfer pipette.
11. For short-term experiments limited to a few hours (e.g., for live analysis of cardiodynamics), embryos can be imaged in dissection media. However, for longer imaging sessions, when embryos are expected to progress through development, the culture medium should be supplemented with rat serum: two parts DMEM/F12, one part rat serum, and 1% 100× penicillin/streptomycin.

12. Place the embryo in a culture dish for recovery in a 5%CO_2 and 37°C incubator for 30 min to 1 h.

3.3 Imaging Setup

Different OCT systems can be utilized for mouse embryo imaging. To acquire images presented in this chapter, we utilized a lab-built spectral-domain OCT system based on the MICRA 5 Titanium-Sapphire laser (Coherent Inc.) with a central wavelength of 800 nm and a bandwidth of ~110 nm [1]. A Fiber-based Michelson interferometer is used for the interference of light reflected from the reference arm and the back-reflected light from the sample arm to form fringes. Interference fringes are spatially resolved using a lab-built spectrometer consisting of a diffraction grating, a focusing lens system, and a line-field CMOS camera (spL4096–140 km, Basler). A 1D depth profile (A-scan) is obtained by performing a fast Fourier transform on fringe data that is transformed into equally sampled data in linear k-space using the laser spectra information. 2D images and 3D volumes are obtained using galvonometer scanners (Cambridge Technology 6220H with 8 mm mirror) mounted on the sample arm over the imaging objective. The system has an A-line rate of ~68 kHz and provides the spatial resolution of ~5 μm (in tissue) in the axial direction and ~4 μm in the transverse direction. The system sensitivity was measured at ~97 dB with a sensitivity drop of ~4 dB over ~1 mm in depth. The OCT sample arm is placed in a commercial environment-controlled incubator during the entire imaging process (Fig. 1b, c) (*see* **Notes 4 and 5**).

3.4 Visualization of Embryonic Structures Using OCT

OCT imaging of live static culture provides visualization of embryonic structural features through early development (*see* **Note 6**). While not in a natural physiological environment, embryos cultured at E7.5–E10.5 stages reach the same developmental hallmarks as embryos in utero [1, 2, 5]. Cellular level resolution allows for visualization and analysis of structural defects in embryonic mouse mutants. As an example, Fig. 2 shows structural phenotypic analysis of embryos at E8.5 and E9.5 with OCT in the Wdr19 mouse model. Wdr19 mutants are embryonic lethal at E10.5 and exhibit defects in neural tube closure and cardiac looping [1]. The figure shows bright-field microscopic images (left) next to 3-D OCT reconstructions (right) of wild-type embryos as controls (Fig. 2a, b) and Wdr19 mutant embryos (Fig. 2c, d) at two developmental stages. The structure of the head region is clearly distinct. In addition to the size difference, the control embryo at E9.5 has a well-closed neural tube at the head region, while the Wdr19 embryo exhibits an open neural tube. In addition, the structure of the heart tube is abnormal in the Wdr19 embryo at E8.5, demonstrating a deficiency in cardiac looping, which is essential for normal cardiogenesis. This defect is clearly seen in the OCT reconstruction.

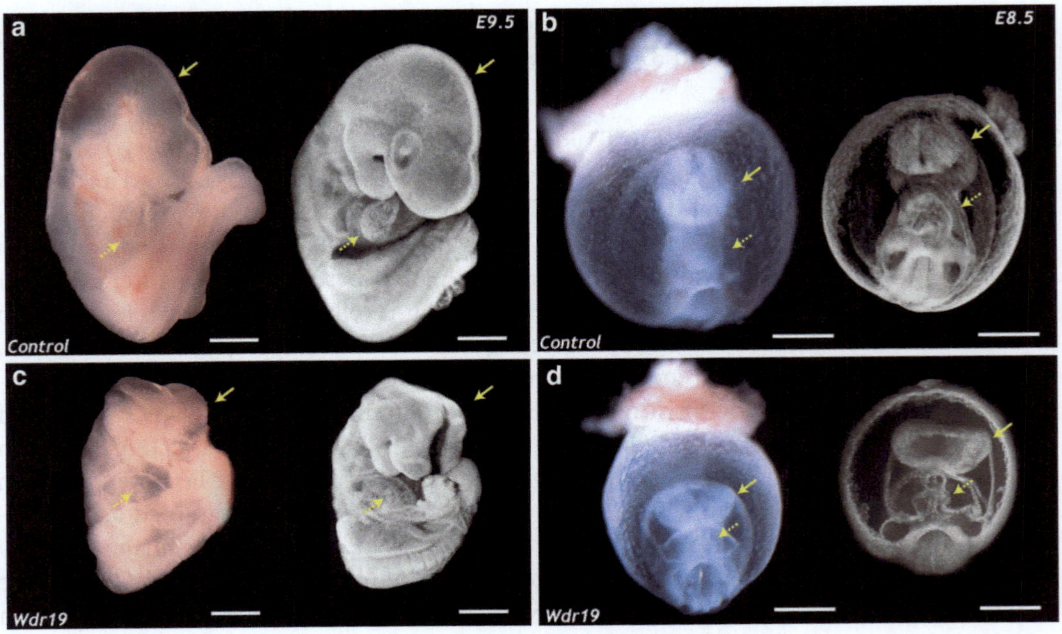

Fig. 2 Structural analysis of early-stage mouse embryos. Brightfield microscopic images (left) and 3D OCT images (right) of control embryos (**a**, **b**) and Wdr19 mutant (**c**, **d**), embryos at E9.5 (**a** + **c**) and E8.5 (**b** + **d**). Solid arrows point at the neural tube in the head region, and dashed arrows point at the heart. Scale bars correspond to 500 μm

For later-stage embryos, the imaging depth of the OCT might be insufficient to visualize internal structures. In this case, embryos can be dissected out of the yolk sac for imaging. While this approach is not compatible with live imaging (since it disrupts circulation), it reveals structural features at the cellular level without application of contrast agents in freshly dissected embryos, which might be beneficial in a variety of applications.

3.5 Cardiodynamic Analysis

A high-volume acquisition rate is critical for cardiodynamic analysis. Nonetheless, traditional OCT systems cannot achieve the ~100 frames per second frame rate required for direct volumetric acquisition and heart wall tracking. To resolve this limitation, a number of algorithms have been proposed to synchronize time sequences acquired over multiple heartbeats at neighboring positions through the entire heart volume [6–14]. These methods rely on the periodic nature of the cardiac cycle. Below is an algorithm based on non-gated synchronization (Fig. 3), which we successfully used for reconstruction of 4-D beating hearts [1, 5, 14]. This approach is best for embryos from E8.5 to E9.5.

1. Position the embryo in the culture dish so that the heart region of the embryo faces up towards the imaging beam. Leaving a piece of deciduum attached to the ectoplacental cone will allow

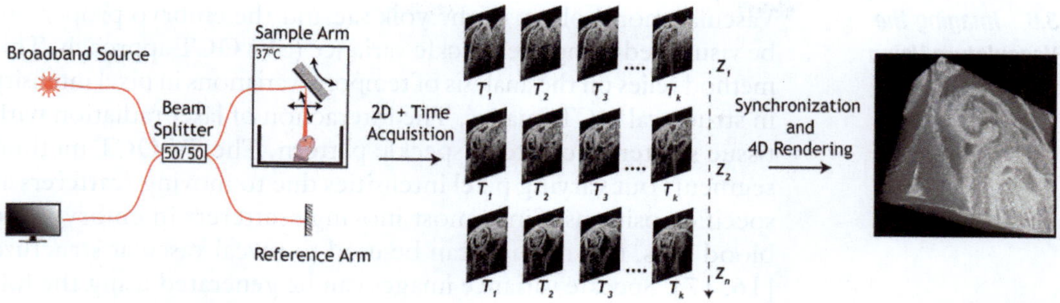

Fig. 3 Volumetric cardiodynamic imaging approach. Adapted from [1]

one to manipulate the tissue into a base and keep the embryo steady.

2. Once the embryo is positioned, spatial acquisition parameters have to be assigned to ensure that the entire volume of heart will be imaged. For E8.5 embryos, our scanning amplitude in the *X* and *Y* direction is ~750 μm ± 150 μm. From this point on, the embryo should remain undisturbed in the incubator for the duration of the imaging session.
3. Make note of the assigned spatial parameters, as they will be necessary for computing the spatial geometry of your sample during post-processing and rendering.
4. Assign scanning parameters. As a suggestion, we assign 600 A-lines and 20,000 B-scans per volume for an E8.5 embryo. The B-scans are continuously acquired at 100 frames per second over a single volume (the *Y*-galvonometer scanner is slowly moving through the volume, while the *X*-galvomometer scanner is quickly moving for B-scan acquisition). As a result, during each heartbeat cycle (the heart rate of the embryo is ~2Hz at this stage), the Y scanner moves by ~2 μm, which is below the lateral resolution of the system.
5. After imaging is completed, OCT data files must be processed into an image sequence of image files corresponding to B-scans for processing and visualization.
6. Define the number of frames per heartbeat cycle. Split the total data set into individual cycles and assign the position for each heartbeat cycle according to the total scan size in the *Y* direction and the total acquired number of heartbeats. We use Imaris v8.0.2 software (Bitplane, Switzerland) for 3D rendering and processing. The heartbeats are synchronized to the same phase of the heartbeat cycle using previously described algorithms [14, 15], revealing the 4-D (3-D plus time) cardiodynamic data set.

3.6 Imaging the Vasculature Using Speckle Variance

Vascular morphology of the yolk sac and the embryo proper can be visualized using the speckle variance (SV) OCT approach. This method relies on the analysis of temporal variations in pixel intensity in structural OCT images. The interaction of laser radiation with tissue scatterers creates a speckle pattern. The SV OCT method segments out varying pixel intensities due to moving scatterers at specific positions. Since most moving scatterers in embryos are blood cells, this method can be used to reveal vascular structure [16, 17]. Speckle variance images can be generated using the following formula, which determines the variance between N images in time, t:

$$SV_{i,j} = \frac{1}{N}\sum_{t=0}^{N-1}\left(I_{i,j,t} - I_{i,j,mean}\right)^2$$

where, $SV_{i,j}$ is the variance at a specific pixel (i,j), $I_{i,j,t}$ is the structural image intensity at that pixel at time t, and $I_{i,j,mean}$ is the mean intensity of pixel (i,j) for all N values [16]. The speckle variance approach provides vascular structure in the developing embryo and extra-embryonic structures. However, movements associated with the beating embryonic heart can degrade the SV signal and limit it to non-moving vascular beds. Using speckle variance and taking advantage of periodic nature of the heart, more advanced image processing algorithms can be implemented to dynamically visualize blood flow within the vasculature and the beating embryonic heart [18].

3.7 Live Imaging of Hemodynamics Using Doppler OCT

Quantitative blood flow analysis is an important component to functionally assess the heart. Doppler OCT relies on the measurement of OCT phase shifts between consecutive A-scans at each pixel and can be used to quantify hemodynamics at the same spatial and temporal resolution as structural OCT imaging [19]. The direction of flow relative to the laser beam needs to be defined for quantitative blood flow analysis. The flow velocity at each pixel can be reconstructed using the formula [20]:

$$v = \frac{\Delta\varphi}{2n\frac{2\pi}{\lambda}\tau\cos\beta}$$

where φ is a Doppler phase shift between adjacent A-scans, n is the refractive index, λ is the central wavelength used for imaging, τ is the time between consecutive A-scans, and β is the angle between the flow direction and the laser beam. The angle β is calculated from structural 2D and 3D data sets acquired from imaged embryos. To image embryonic mice, we usually assume a refractive index of $n = 1.4$.

The resolution of Doppler OCT allows for quantitative hemodynamic analysis from bulk blood cell movement [21] as well as individual circulating erythrocytes at early circulation stages when most blood cells are still restricted from flowing in the blood islands [22]. Spatially and temporally resolved blood flow profiles can be reconstructed from these measurements. Hemodynamic measurements acquired by Doppler OCT in cultured mouse embryos are comparable to measurements made using fast-scanning confocal microscopy in superficial vessels. However, flow measurements deep within the embryo are not possible with confocal microscopy due to the imaging depth limitation of this method, and unique to the Doppler OCT technique [21, 23].

Volumetric blood flow analysis in the beating heart with Doppler OCT can be performed using the synchronization approach described above for structural imaging [14, 24]. For that, structural and Doppler OCT images are exported and treated as two channels of the same data set. The synchronization is performed based on the structural data set, while the Doppler OCT frames are passively rearranged and registered based on the corresponding structural data arrangements. Figure 4 shows an example of volumetric Doppler OCT analysis of blood flow in the embryonic heart. The blood flow labeled with blue (toward the detector) and magenta (away from the detector) is clearly distinguishable in the heart and extra-embryonic vessels at different phases of the heartbeat cycle.

Analysis of absolute blood flow velocity in the heart might be challenging due to the difficulty of evaluating the flow direction. At early heart tube stages, when the flow is laminar and the direction of flow is obvious based on the 3D structure of the heart tube, the absolute flow velocity profiles in the heart can be reconstructed and are comparable with ultrasonic measurements acquired in utero [24]. This provides an opportunity for detailed quantitative volumetric biomechanical analysis in the embryonic heart by correlating heart wall dynamics with local flow measurement [3, 24]. An example of correlative analysis of the cardiodynamics and hemodynamics is shown in Fig. 5. This can reveal the details of the cardiac cycle

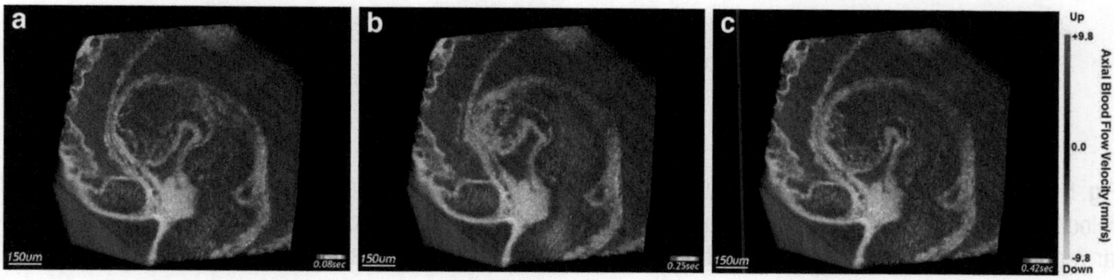

Fig. 4 Volumetric hemodynamic imaging in the embryonic heart. Snapshots from 4D hemodynamic imaging of a late E8.5 wildtype embryonic mouse heart at different phases of the cardiac cycle. Panels a,b, and c represent different phases of the cardiac cycle. The structural data is shown in gray scale while the Doppler OCT velocity measurement is overlaid in color

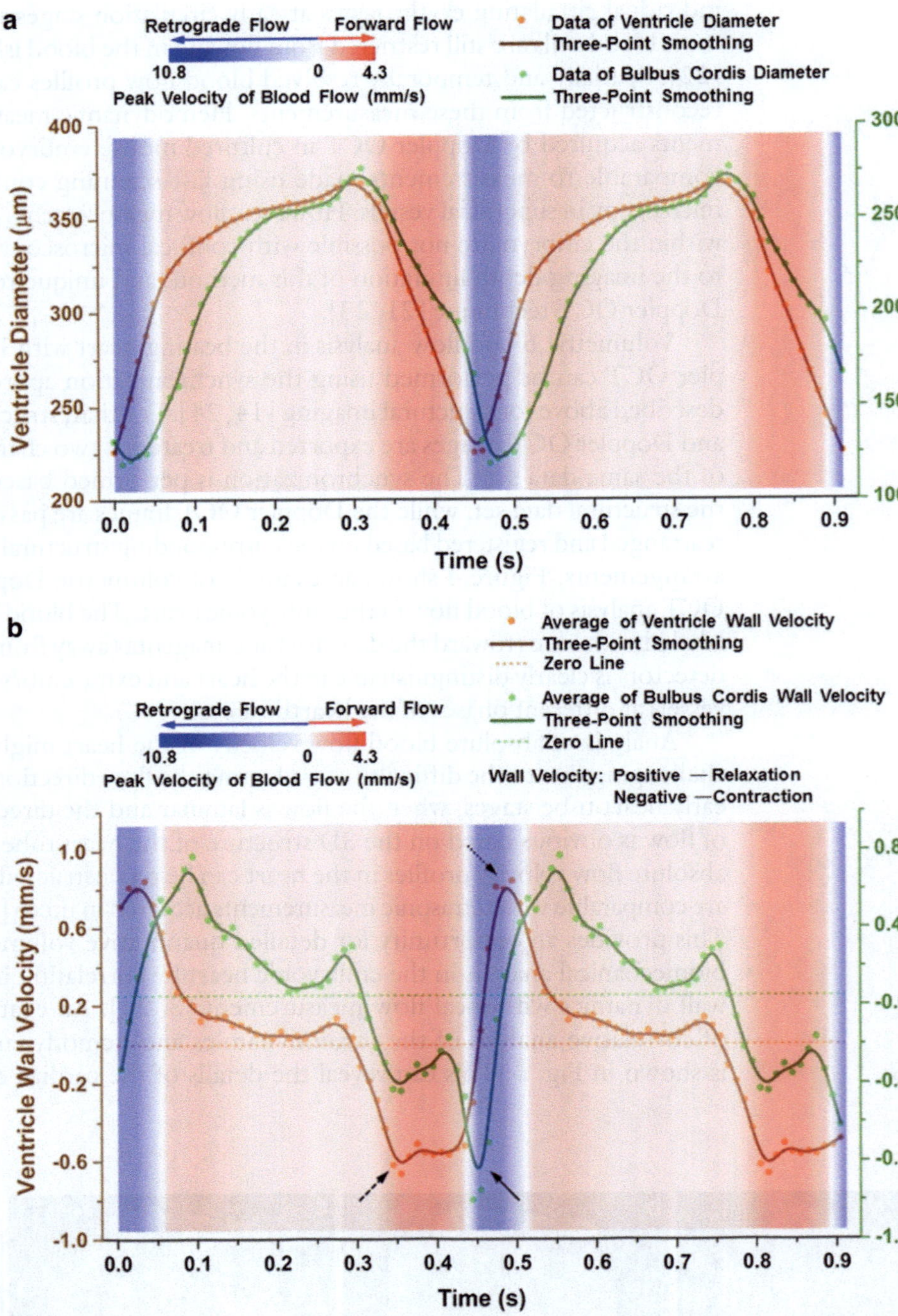

Fig. 5 Analysis of hemodynamics in relation to heart wall dynamics. (**a**) Diameters and (**b**) average heart wall velocities of the ventricle and the bulbus cordis are plotted with blood flow dynamics in the bulboventricular region of a wildtype E9.0 embryo. The heartbeat is duplicated for a second cycle for better visualization. Adapted from [5]

previously not known and opens a door for studies on the biomechanical regulation of cardiac function.

Here, we describe methods for dynamic imaging of the developing embryonic mouse heart. Using OCT and postprocessing methods, one can extract detailed structural and functional information. This is highly relevant when trying to understand how dynamic and mechanical changes influence development and result in mutant phenotypes. Combined with molecular genetic approaches, these methods can help us generate a more complete view of all the parameters that regulate development.

4 Notes

1. Rat serum is available commercially; however, we prepare our own serum via the protocol provided here.
2. We use ether for rat serum extraction because it evaporates from the serum, unlike other anesthetics.
3. We find it helpful to change out the media during the dissection to ensure pH and temperature are consistent.
4. We recommend tilting the scanning head slightly (~5°) to minimize light reflection from the media surface and the dish bottom, which might significantly improve the imaging quality.
5. We recommend removing the scanning head from the incubator when not in use or turning off the incubator and removing the water tray from the stage to minimize the effect of moisture on the scanning head's electronics and mechanics.
6. It is essential to maintain 37°C throughout the imaging session for the imaging stage, the culture media, and the dissecting stage. Temperature variations can have a dramatic effect on cardiodynamics and embryo viability.

Acknowledgments

This work was supported by the National Institute of Health with grants R01HL120140, U54HG006348, R01HD086765, and T32HL07676, and by the Optical Imaging and Vital Microscopy Core at Baylor College of Medicine.

Conflict of Interest The authors have no conflicts of interest to declare that are relevant to the content of this chapter.

References

1. Lopez III AL, Wang S, Larin KV, Overbeek PA, Larina IV (2015) Live four-dimensional optical coherence tomography reveals embryonic cardiac phenotype in mouse mutant. J Biomed Opt 20(9):90501
2. Wang S, Garcia MD, Lopez AL, Overbeek PA, Larin KV, Larina IV (2017) Dynamic imaging and quantitative analysis of cranial neural tube closure in the mouseembryo using optical coherence tomography. Biomed Opt Express 8(1):407–419
3. Wang S, Larina I (2020) Live mechanistic assessment of localized cardiac pumping in mammalian tubular embryonic heart. J Biomed Opt Centers for Disease Control and Prevention (CDC) 25(8):1–19
4. Wang S, Larina IV (2022) Following the beat: imaging the valveless pumping function in the early embryonic heart. J Cardiovasc Dev Dis 9(8):267
5. Wang S, Lakomy DS, Garcia MD, Lopez AL, Larin KV, Larina IV (2016) Four-dimensional live imaging of hemodynamics in mammalian embryonic heart with Doppler optical coherence tomography. J Biophotonics 9(8):837–847
6. Liu A, Wang R, Thornburg KL, Rugonyi S (2009) Efficient postacquisition synchronization of 4-D nongated cardiac images obtained from optical coherence tomography: application to 4-D reconstruction of the chick embryonic heart. J Biomed Opt 14(4):044020
7. Jenkins MW, Rothenberg F, Roy D, Nikolski VP, Hu Z, Watanabe M, Wilson DL, Efimov IR, Rollins AM (2006) 4D embryonic cardiography using gated optical coherence tomography. Opt Express 14(2):736–748
8. Jenkins MW, Chughtai OQ, Basavanhally AN, Watanabe M, Rollins AM (2007) In vivo gated 4D imaging of the embryonic heart using optical coherence tomography. J Biomed Opt 12(3):030505
9. Mariampillai A, Standish BA, Munce NR, Randall C, Liu G, Jiang JY, Cable AE, Vitkin IA, Yang VX (2007) Doppler optical cardiogram gated 2D color flow imaging at 1000 fps and 4D in vivo visualization of embryonic heart at 45 fps on a swept source OCT system. Opt Express 15(4):1627–1638
10. Larin KV, Larina IV, Liebling M, Dickinson ME (2009) Live imaging of early developmental processes in mammalian embryos with optical coherence tomography. J Innov Opt Health Sci 2(3):253–259
11. Gargesha M, Jenkins MW, Wilson DL, Rollins AM (2009) High temporal resolution OCT using image-based retrospective gating. Opt Express 17(13):10786–10799
12. Daniels K, Daugherty J, Jones J, Mosher W (2015) Current contraceptive use and variation by selected characteristics among women aged 15–44: United States, 2011–2013. Natl Health Stat Rep 86:1–14
13. Division of Reproductive Health, National Center for Chronic Disease Prevention and Health Promotion, Centers for Disease Control and Prevention (CDC) (2013) 2013 U.S. Selected practice recommendations for contraceptive use, 2013: adapted from the World Health Organization selected practice recommendations for contraceptive use, 2nd edn. Morbidity and Mortality Weekly Report (MMWR), vol. 62(RR-05). Centers for Disease Control and Prevention (CDC). pp 1–60
14. Faubert AC, Larina IV, Wang S (2022) Open-source, highly efficient, post-acquisition synchronization for 4D dual-contrast imaging of the mouse embryonic heart over development with optical coherence tomography. Biomed Opt Express 14(1):163–181
15. Liebling M, Forouhar AS, Gharib M, Fraser SE, Dickinson ME (2005) Four-dimensional cardiac imaging in living embryos via postacquisition synchronization of nongated slice sequences. J Biomed Opt 10(5):054001
16. Sudheendran N, Syed SH, Dickinson ME, Larina IV, Larin KV (2011) Speckle variance OCT imaging of the vasculature in live mammalian embryos. Laser Phys Lett 8(3):247–252
17. Mariampillai A, Leung MK, Jarvi M, Standish BA, Lee K, Wilson BC, Vitkin A, Yang VX (2010) Optimized speckle variance OCT imaging of microvasculature. Opt Lett 35(8):1257–1259
18. Grishina OA, Wang S, Larina IV (2017) Speckle variance optical coherence tomography of blood flow in the beating mouse embryonic heart. J Biophotonics 10(5):735–743
19. Chen Z, Milner TE, Srinivas S, Wang X, Malekafzali A, van Gemert MJ, Nelson JS (1997) Noninvasive imaging of in vivo blood flow velocity using optical Doppler tomography. Opt Lett 22(14):1119–1121
20. Vakoc B, Yun S, de Boer J, Tearney G, Bouma B (2005) Phase-resolved optical frequency domain imaging. Opt Express 13(14):5483–5493

21. Larina IV, Sudheendran N, Ghosn M, Jiang J, Cable A, Larin KV, Dickinson ME (2008) Live imaging of blood flow in mammalian embryos using Doppler swept-source optical coherence tomography. J Biomed Opt 13(6):060506
22. Larina IV, Ivers S, Syed S, Dickinson ME, Larin KV (2009) Hemodynamic measurements from individual blood cells in early mammalian embryos with Doppler swept source OCT. Opt Lett 34(7):986–988
23. Jones EA, Baron MH, Fraser SE, Dickinson ME (2004) Measuring hemodynamic changes during mammalian development. Am J Physiol Heart Circ Physiol 287(4):H1561–H1569
24. Wang S, Lakomy DS, Garcia MD, Lopez III AL, Larin KV, Larina IV (2016) Four-dimensional live imaging of hemodynamics in mammalian embryonic heart with Doppler optical coherence tomography. J Biophotonics 9(8):837–847

Chapter 3

3D Imaging of Gene Expression Domains in the Mouse Heart

Wataru Katano, Yuki Tajika and Kazuko Koshiba-Takeuchi

Abstract

In situ hybridization (ISH) is a powerful method for analyzing gene expression patterns in tissue sections and whole-mount specimens. Three-dimensional (3D) imaging of ISH signals helps to understand the gene expression patterns in the complex structure of developing organs. This chapter describes the protocol for creating 3D images of ISH signals in a cost-effective and time-efficient manner using the correlative microscopy and block-face imaging (CoMBI) system. A unique feature of the CoMBI system is the ability to obtain both serial block-face images and sections as needed for preferential staining, such as hematoxylin and eosin staining. Correlating section images and 3D data helps to annotate gene expression patterns and understand the anatomy of developing organs. We describe the protocol for the sample preparation (ISH and embedding), acquisition of serial block-face images, and image processing to create 3D images.

Keywords In situ hybridization, Whole-mount, Gene expression, CoMBI, 3D imaging

1 Introduction

In situ hybridization (ISH) is one of the most useful methods for analyzing gene expression patterns in tissue sections and whole-mount specimens, which are traditionally represented as two-dimensional (2D) microscopic images of sections, or macroscopic surface views of the whole specimen [1]. Obtaining three-dimensional (3D) gene expression patterns by ISH helps to understand complex structures in the developing organ. Conventionally, researchers create 3D images of gene expression patterns by performing ISH on numerous serial sections, photographing them individually, and reconstructing them into a 3D image with considerable time and labor [2, 3]. This conventional method is useful for understanding 3D gene expression patterns, but it is difficult to create a smooth 3D image due to the deformation of sections and the failure to collect sections due to human error. Recently, we have reported a cost-effective and time-efficient approach to represent ISH signals in 3D using correlative microscopy and block-face imaging (CoMBI)

Paul Delgado-Olguín (ed.), *Mouse Embryogenesis*, Methods in Molecular Biology, vol. 3010,
https://doi.org/10.1007/978-1-0716-5126-1_3,

[4]. The CoMBI system automatically acquires serial block-face images from paraffin-embedded or frozen blocks using a consumer digital camera, macro lens, microtome, and self-made camera shutter controller [5–7]. The block-face captured using the CoMBI system showed the internal structures of the specimen in a natural feature and color with a high signal-to-noise ratio. In addition, sections can be taken as needed during serial block-face imaging for preferential staining, such as hematoxylin and eosin staining. Obtaining both serial block-face images and some sections could be useful for annotating the site of ISH signals. The outline of the combined method of ISH and CoMBI is as follows: (1) whole-mount ISH on mouse embryos, (2) preparation of the sample embedded blocks as paraffin or frozen blocks, (3) application of the CoMBI system to obtain a series of block-face images, and (4) image processing to create 3D images (Fig. 1).

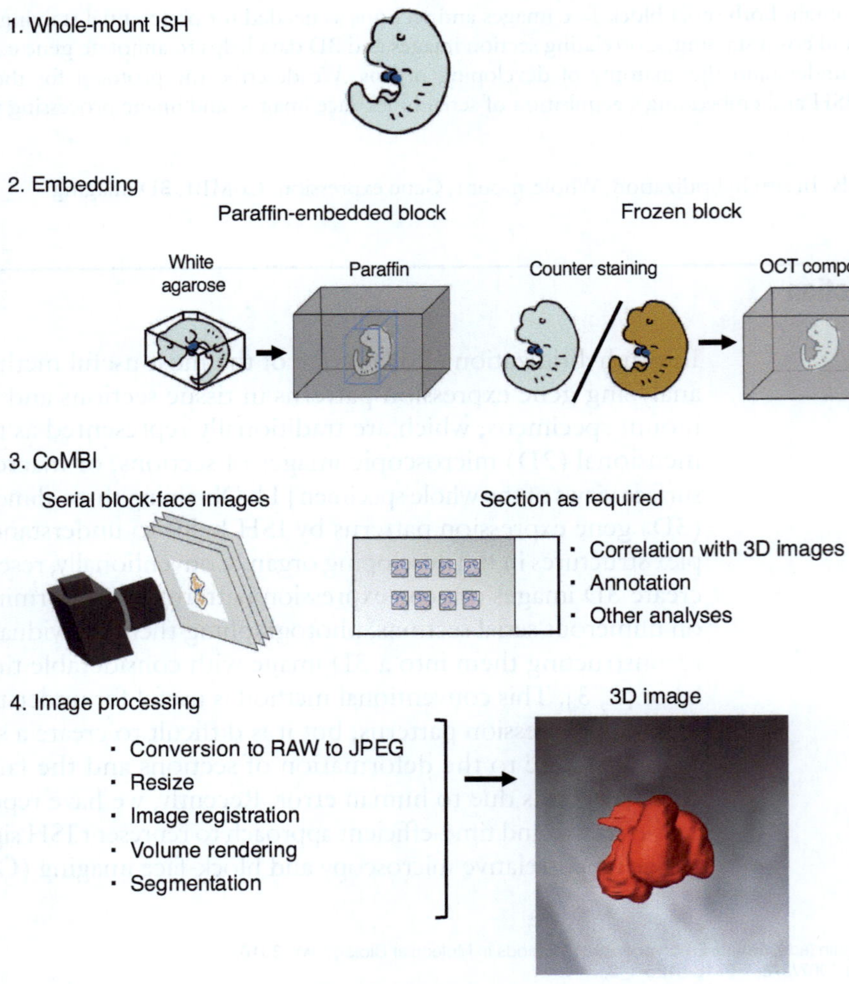

Fig. 1 Schematic diagram of the CoMBI analysis for ISH-labeled specimens

2 Materials

2.1 Whole-Mount In Situ Hybridization (WISH)

1. RNA probe (*see* **Note 1**).
2. PBS: Phosphate-buffered saline.
3. 4% w/v paraformaldehyde (PFA) in PBS (*see* **Note 2**).
4. PBST: PBS containing 0.1% Tween-20.
5. Hybridization buffer: 50% formamide (deionized), 5× SSC pH 4.5, 50 μL/mL yeast tRNA, 1% SDS (filtrated), 50 μg/mL heparin, RNase-free water (store at −20°C).
6. Washing solution 1: 50% formamide, 5× SSC [pH 4.5], 1% SDS in RNase-free water (store at 4°C).
7. Washing solution 2: 50% formamide, 2× SSC [pH 4.5] in RNase-free water (store at 4°C).
8. TBST: Tris-buffered saline containing 0.1% Tween-20.
9. 10% sheep serum in TBST (store at −20°C).
10. NTMT buffer: 100 mmol/L NaCl, 50 mmol/L $MgCl_2$, 100 mmol/L Tris-HCl (pH 9.5), 0.1% Tween20, and 2 mmol/L levamisol.
11. BM purple AP substrate.
12. Filters (0.22 μm-pore size) for removing insoluble debris in PBS, TBST, and NTMT buffer.

2.2 Frozen Block

1. 1% w/v tannic acid in PBS (*see* **Note 3**).
2. 10% w/v sucrose in PBS.
3. 30% w/v sucrose in PBS.
4. White OCT compound: OCT compound containing 5%–10% white watercolor paste (*see* **Note 4**).
5. Embedding molds made of acrylic plate (*see* **Note 5**).

2.3 Paraffin Block

1. White agarose gel: Dissolve 500–1000 mg of white watercolor paste and 100 mg of fish gelatin in 10 mL of distilled water, which can be stored at −20°C, and then mix with 1% w/v low-melting-point agarose at a ratio of 1:4.

2.4 CoMBI System

CoMBI devices synchronize the microtome handles' position and the camera shutter release timing to allow the automatic acquisition of serial block-face images. The CoMBI system comprises a self-made camera shutter control device, imaging devices, and general microtomes (Fig. 2). All information, including components, code, and circuit for constructing CoMBI devices, are openly available

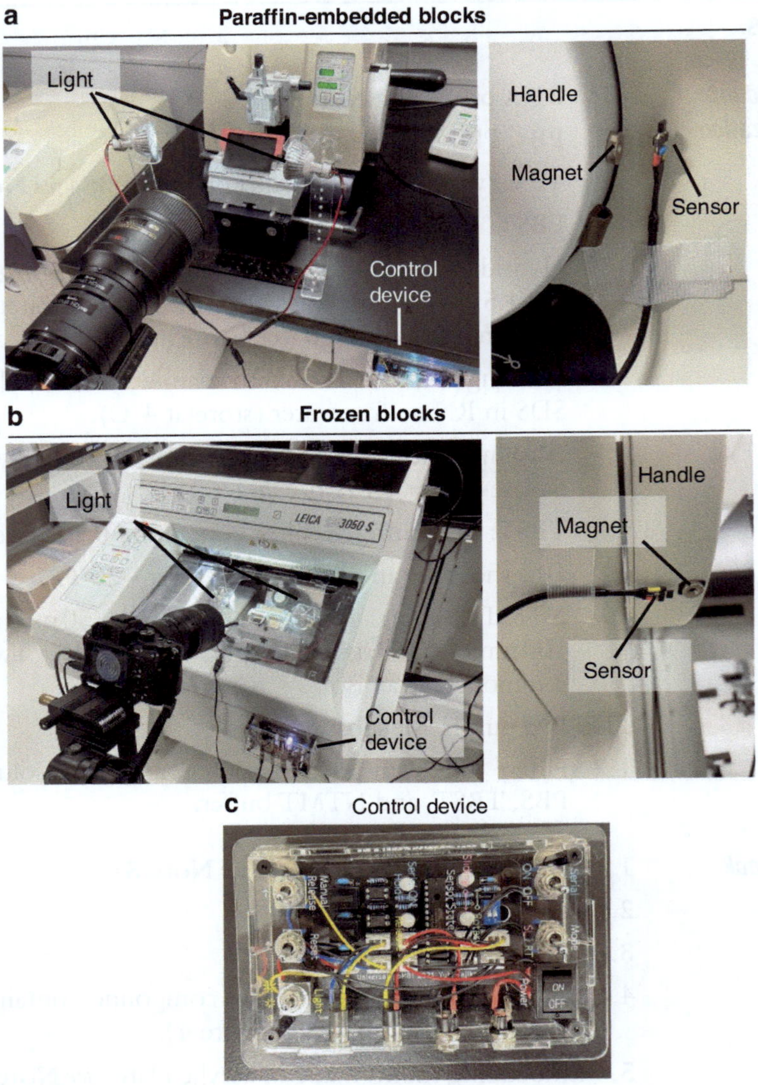

Fig. 2 CoMBI system. (**a**) CoMBI system for analyzing paraffin-embedded specimens. (**b**) CoMBI system for frozen specimens. (**c**) Camera shutter control devices receive signals from a magnetic sensor

at GitHub (https://github.com/combi-3d), and the instruction manual is also available (https://combi-3d.github.io/manual/).

1. Imaging device: a digital camera, a macro lens, a mount adaptor, a teleconverter, a tripod, and a geared head.
2. Camera shutter control device: a self-made device using a microcontroller, an optocoupler, a piece of magnet, and a magnetic sensor.
3. Illumination device: two 3W LED lamps.

3 Methods

3.1 Setting of CoMBI System into a Microtome

1. Set the camera in front of the microtome and direct the camera perpendicular to the block-face (Fig. 2, left column).
2. Set LED lamps on both sides of the block (Fig. 2, left column).
3. Attach a magnet to a microtome handle and determine the sensor position (Fig. 2, right column).

3.2 Whole-Mount In Situ Hybridization of Mouse Embryos

1. Remove the yolk sac and pericardium of the mouse embryo in PBS (*see* **Note 6**).
2. Fix embryos in 4% w/v PFA/PBS at 4°C overnight.
3. Wash embryos in PBST twice for 5 min.
4. Dehydrate embryos with MeOH/PBST series. Incubate in 25% MeOH/PBST, 50% MeOH/PBST, and 75% MeOH/PBST for 10 min at RT, respectively. Then incubate in 100% MeOH twice for 30 min at RT.
5. Wash embryos in PBST twice for 5 min.
6. Treat embryos with 5–20 μg/mL Proteinase K in PBST for 30 min at RT (*see* **Note 7**).
7. Incubate embryos in 2 mg/mL glycine in PBST for 15 min.
8. Wash embryos in PBST twice for 5 min.
9. Fix embryos with 4% w/v PFA/PBST with 0.2% glutaraldehyde for 20 min at RT.
10. Wash embryos in PBST twice for 5 min.
11. Incubate embryos in hybridization buffer without the RNA probe at 65°C for 1 h.
12. Add the RNA probe to achieve a final concentration of 1:100 and then hybridize at 65°C overnight.
13. Remove the hybridization buffer with the RNA probe. The used RNA probe can be reused several times if stored at −20°C.
14. Wash embryos in wash solution 1 three times for 30 min at 65°C.
15. Wash embryos in washing solution 2 three times for 30 min at 65°C.
16. Wash embryos in washing solution 2/TBST (1:1) for 10 min at 6°C.
17. Wash embryos in TBST three times for 10 min at RT.
18. Incubate embryos in a blocking solution for 1 h at RT.
19. Incubate embryos in anti-DIG antibody (1:2500) in a blocking solution at 4°C overnight.

20. Wash embryos with TBST three times for 5 min at RT.
21. Wash embryos with TBST six times for 5 min at RT.
22. Wash embryos with NTMT buffer three times for 10 min at RT.
23. Incubate embryos in BM purple. During the reaction, put the samples in a dark place to avoid lights.
24. Check the signal after 15 min of incubation, and then every 30 min thereafter.
25. Wash embryos with PBS.
26. Fix embryos with 4% w/v PFA/PBS. The samples can be stored at 4°C.

3.3 Preparing Paraffin Blocks and Imaging Using the CoMBI System

1. Wash specimens with PBS twice for 5 min at RT and acquire images before embedding (Fig. 3a).
2. Immerse specimens in the desired orientation in white agarose gel and incubate at 4°C for 30 min (*see* **Note 8**).
3. Fix white agarose blocks with 4% PFA/PBS at 4°C overnight to harden the blocks (Fig. 3b).

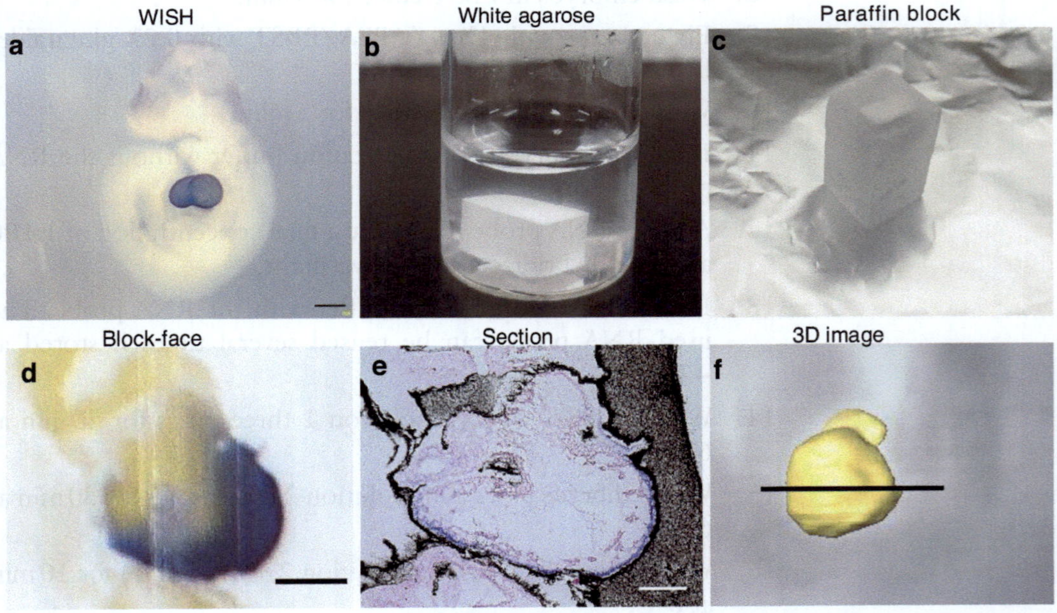

Fig. 3 CoMBI system for paraffin-embedded specimens. (**a**) Whole mount ISH was performed on mouse embryos. The developing heart was labeled with a *Myl2* RNA probe. (**b**, **c**) The embryo was embedded in white agarose, and then embedded in paraffin. (**d**) Serial block-face images were acquired by the CoMBI system. One of the serial images is shown as an example. (**e**) Sections were collected as needed and stained with H&E. (**f**) A 3D image was reconstructed from serial block-face images. Segmented heart (yellow) and volume-rendered embryo body (white) are shown. Scale bars: 500 μm (**a**), 200 μm (**d**), 100 μm (**e**)

4. Dehydrate white agarose blocks with MeOH/PBST series. Incubate in 25% MeOH/PBST, 50% MeOH/PBST, and 75% MeOH/PBST for 1 h at RT, respectively. Then incubate in 100% MeOH three times for 1 h at RT (*see* **Note 9**).
5. Incubate white agarose blocks in xylene twice for 1 h at RT.
6. Incubate white agarose blocks in xylene/paraffin (1:1) for 1 h at 62°C (*see* **Note 10**)
7. Incubate white agarose blocks in paraffin twice for 1 h at 62°C.
8. Immerse white agarose blocks in the desired orientation in the paraffin solution and cool on ice (Fig. 3c).
9. Cut out as a paraffin block. The block contains the mouse embryo embedded in white agarose and has an adequate size to be set in the microtome.
10. Slice the block to create a flat surface. Set the camera focus carefully. Some cameras have useful functions for focusing, such as magnified display and focus peaking.
11. Acquire a ruler image for the scale measurement.
12. Slice the block and acquire serial block-face images (Fig. 3d). Collect sections as needed (Fig. 3e). The thickness is usually set at 5–10 μm in our laboratory. Float the collected sections on Milli-Q water, and then dry them overnight at 42°C.

3.4 Preparing Frozen Blocks and Imaging Using the CoMBI System

1. Wash specimens with PBS twice for 5 min at RT.
2. Treat specimens with 1% w/v tannic acid/PBS for 1–2 h for counterstaining (Fig. 4a, b, *see* **Note 11**).
3. Wash specimens with PBS twice for 5 min at RT.

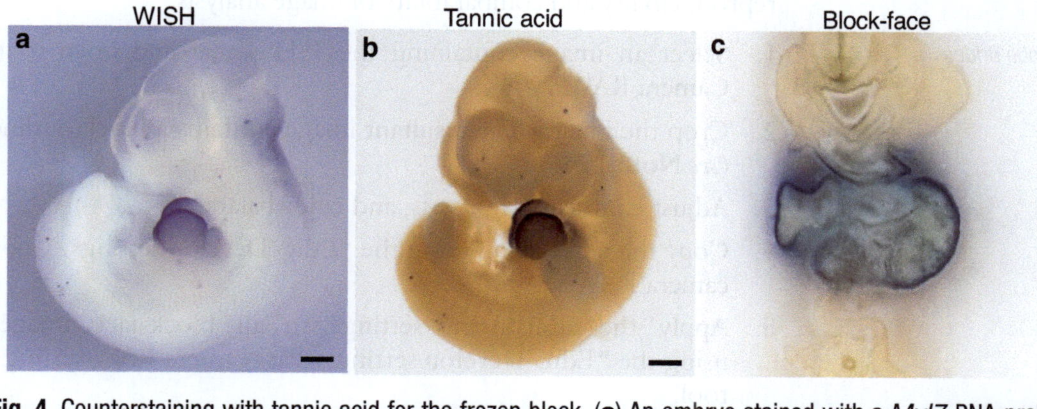

Fig. 4 Counterstaining with tannic acid for the frozen block. (**a**) An embryo stained with a *Myl7* RNA probe by WISH. (**b**) The embryos were counterstained with 1% tannic acid. This staining provides better visualization of the internal structures of the embryo in the block face. (**c**) The embryo in (**b**) was frozen and proceeded to block-face imaging. Tannic acid is effective in increasing the visibility of the internal structures of the embryo. Scale bars: 1 mm

4. Incubate specimens in 10% w/v sucrose/PBS for 30 min at 4°C.
5. Incubate specimens in 30% sucrose/PBS overnight at 4°C. Specimens can be stored at 4°C for a week (*see* **Note 12**).
6. Cover specimens with white OCT compound (*see* **Note 13**).
7. Immerse specimens in the desired orientation in the OCT compound without white watercolor paste (*see* **Note 14**) and freeze quickly with liquid nitrogen, ensuring the freezing proceeds from the bottom to the top (*see* **Note 15**). The frozen blocks can be stored at −80°C until use.
8. Place the frozen block in a cryostat chamber at −20°C and leave for at least 10 min until the temperature of the chamber and block are the same.
9. Slice the block to create a flat surface. Set the camera focus carefully. Some cameras have useful functions for focusing, such as magnified display and focus peaking.
10. Acquire serial block face images (Fig. 4c) and collect sections as needed. The thickness is usually set at 8–20 μm in our laboratory.

3.5 Adjusting the RAW Image and Converting into the JPEG Image

Serial block-face images are saved as RAW images with 12-bit or 14-bit gradation. We convert the images to 8-bit images such as JPEGs and TIFFs using Adobe Camera RAW, which runs in Bridge and Lightroom (*see* **Note 16**). The protocol here also shows the case of Sony Imaging Edge. We prefer to use Adobe Camera RAW because it is better at adjusting the intensity of a specific color and enhancing low-contrast boundaries (Fig. 5). Sony's app is free for Sony camera users, which gives it a cost advantage. These applications all allow you to save all parameters for image adjusting, ensuring the reproducibility and comparability of image analysis.

Adobe Bridge

1. Select an image containing the ISH signal and open it in Camera RAW.
2. Crop the image. The resultant image contains a block outline (*see* **Note 17**).
3. Adjust contrast, brightness, and color balance.
4. Copy the edit setting using the "Edit/Develop settings/Copy camera RAW setting" tool.
5. Apply the adjustment settings to all block-face images using the "Edit/Develop settings/Past camera RAW setting" tool.
6. Select all images and open them in Camera RAW.
7. Convert the image format from RAW to JPEG (*see* **Note 18**).

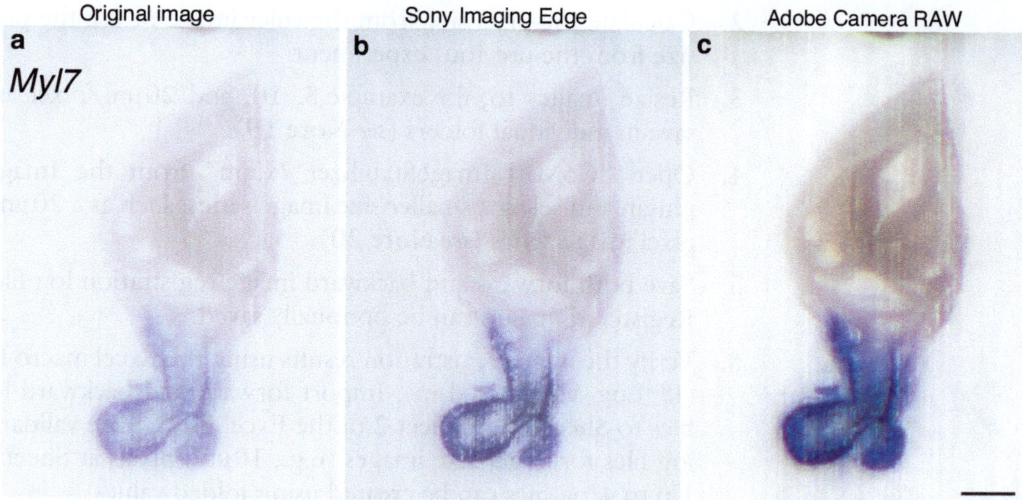

Fig. 5 Difference of RAW image adjustment capability between applications. (**a**) An original image of the paraffin block-face of the embryo stained with a *Myl2* RNA probe (blue). (**b**, **c**) Adjusted images using Sony Imaging Edge or Adobe Camera RAW. Both applications improve the visibility of the ISH signals. Adobe Camera RAW improves the contrast between the embryo body and the surrounding agarose. Scale bar: 250 μm

Lightroom

1. Select an image containing the ISH signal.
2. Crop the image. The resultant image contains a block outline (*see* **Note 17**).
3. Adjust contrast, brightness, and color balance.
4. Apply the adjustment settings to all images using the "Synchronize" tool. The settings can be saved as presets.
5. Export as JPEG image (*see* **Note 18**).

Imaging Edge

1. Open an image file from "Edit" and select a specific image.
2. Adjust the contrast and brightness, then save the adjustment settings file.
3. Open a file from "Viewer" and select all images.
4. Convert image type from RAW to JPEG, applying the adjustment settings to all images using the "Export" tool (*see* **Note 18**).

3.6 Resizing, Crop, and Image Registration

Resizing and image registration can be performed automatically by ImageJ macros and Excel VBA, which are available on GitHub (https://github.com/combi-3d/ImageJ). Image registration (Steps 4–7, below) can also be done using Amira Software (a paid application).

1. Open "CoMBI_resize.ijm" from the ImageJ plugin and select a folder containing serial JPEG images obtained by CoMBI.

2. Calculate the pixel size from the ruler image or use the pixel size from the previous experiment.
3. Resize images to, for example,5, 10, and 20 μm/pixel and save in individual folders (*see* **Note 19**).
4. Open "CoMBI_ImageStabilizer_2x.ijm" from the ImageJ plugin and select a smaller size image series, such as a 20 μm/pixel image series (*see* **Note 20**).
5. Save both forward and backward image registration log files. Registered images can be optionally saved.
6. Verify the image registration results using the Excel macro file (IS_Log_Validator.xlsm). Import forward and backward log files to Sheet 1 and Sheet 2 of the Excel file. Create validated log files for larger size images (e.g., 10 μm/pixel) at Sheet 4. Up to 4 log files can be created using folded values.
7. Open "CoMBI_Crop1_ISApplier_Crop2_SelectLog.ijm" from the ImageJ plugin and input pixel size of serial images, then select a validated log file and a folder containing serial images to be registered (*see* **Note 21**).
8. Crop images to include the sample area with enough margins to reduce the image size.
9. Image registration by "Image Stabilizer Log Applier."
10. Crop again to include the sample area with a minimal margin.
11. Create grayscale images for volume rendering.

3.7 3D Reconstruction

We introduce the protocol for Amira and 3D Slicer for 3D reconstruction. Both applications have a lot of functions, though we focused on essential protocols for 3D reconstruction and volume measurement of ISH signals.

Amira

1. Import grayscale images into Amira. Put the voxel size with X and Y as the length per pixel and Z as the section thickness, respectively.
2. Adjust misaligned images using the "Alignslice" tool.
3. Use the "Volren" tool for the volume rendering.
4. Use the "LabelFields/Labeling" tool for the segmentation of ISH signals.
5. Quantify the volume of ISH signals using the "LabelFields/Measure" tool in Amira for statistical analysis.

3D Slicer

1. Import grayscale and color image folders into 3D Slicer using "File/Add data/Choose Directory Add."
2. Use the "Modules/Volumes/Volume rendering" tool to perform the volume rendering.

3. Use the "Modules/Volumes/Segmentation" tool to perform the segmentation.
4. Quantify the volume of ISH signals using the "Modules/Quantification/Segment status" tool for statistical analysis.

4 Notes

1. Design RNA probes longer than 500 bp. Short RNA probes less than 200 bp would not detect specific signals.
2. Use fresh 4% w/v PFA/PBS. The old PFA solution increases the background signal.
3. Use fresh 1% w/v tannic acid/PBS. The tannic acid solution will become dark brown over time.
4. The white color paste and OCT compound will separate within a few days. Mix gently before use.
5. Acrylic molds are recommended because they create a cuboid block to give the identical block outline in block-face images. Molds can be made using laboratory tubes (1.5, 15, and 50 mL), which are cut into pieces. To reduce the deformative effect of ice expansion during freezing, one or two splits are needed in the wall of the mold.
6. If the embryonic day is later than 11.5, the RNA probe cannot penetrate deep tissue, such as the dorsal region of the atria. We recommend dissecting the specific tissue to improve probe penetration.
7. Proteinase K concentration varies with embryonic stages. Treatment at high concentrations increases the ISH signal but decreases the integrity of the specimens.
8. The opacification with white agarose improves the visibility of the surface of the specimen and makes it easier to embed the specimen in the desired orientation during paraffin embedding.
9. The incubation time of white agarose blocks for dehydration depends on the block size.
10. Keep below 62°C for incubation of the white agarose block in paraffin solution. High temperatures (around 70°C) will cause the agarose to melt.
11. Validate the reaction duration by observing the specimen under staining. Staining with tannic acid enhances the contrast between the specimen and the OCT compound and improves the visibility of the internal structures of the specimen.

12. Store the specimen stained with tannic acid in a sucrose solution, not PBS. Specimens in PBS become dark brown over time.
13. Opacification of the OCT compound with white watercolor paste improves the visibility of the tissue surface.
14. Using the OCT compound without white watercolor paste makes it easier to immerse the specimen in the desired orientation.
15. Freezing the OCT compound in a freezer may cause the deformation of the specimen, as slower freezing forms more ice crystals.
16. JPEG is a lossy compressed format, and its data size is small, while TIFF is an uncompressed format and is generally used in laboratories, but requires more storage space. Choose the appropriate image format for your experiment and environment, taking into account reproducibility and comparability.
17. The block outline is required for image registration in ImageJ. When image registration is performed using Amira software, the block outline is not required.
18. The conversion process needs 1GB of RAM per 100 images. Overloading the computer memory will result in slow conversion speed. We are using a computer with at least 16 GB of RAM.
19. ImageJ macro includes creating new folders to save resized images.
20. Create custom-resized images based on sample size. By default, 5, 10, and 20 μm/pixel folders are created. Custom pixel sizes can also be selected.
21. ImageJ v1.53 s or newer is recommended. There is a difference between ImageJ v1 and FIJI/ImageJ when opening text files. CoMBI_Crop1_ISApplier_Crop2_SelectLog.ijm for ImageJ v1 on Windows and Mac. CoMBI_Crop1_ISApplier_Crop2_CopyLog.ijm for FIJI on Mac.

Acknowledgments

This work was supported by The Inoue Enryo Memorial Grant, TOYO University (K.K.-T.).

Conflict of Interest The authors have no conflicts of interest to declare that are relevant to the content of this chapter.

References

1. Koshiba-Takeuchi K (2018) Whole-mount and section in situ hybridization in mouse embryos for detecting mRNA expression and localization. Methods Mol Biol 1752:123–131
2. Soufan AT, Ruijter JM, Van Den Hoff MJ, de Boer PA, Hagoort J, Moorman AF (2003) Three-dimensional reconstruction of gene expression patterns during cardiac development. Physiol Genomics 13(3):187–195
3. Soufan AT, Van Den Hoff MJ, Ruijter JM, de Boer PA, Hagoort J, Moorman AF (2004) Reconstruction of the patterns of gene expression in the developing mouse heart reveals an architectural arrangement that facilitates the understanding of atrial malformations and arrhythmias. Circ Res 95(12):1207–1215
4. Sutrisno AA, Katano W, Kawamura H, Tajika Y, Koshiba-Takeuchi K (2023) Combined method of whole mount and block-face imaging: acquisition of 3D data of gene expression pattern from conventional in situ hybridization. Dev Growth Differ 65(1):56–64
5. Tajika Y, Murakami T, Iijima K, Gotoh H, Takahashi-Ikezawa M, Ueno H, Yoshimoto Y, Yorifuji H (2017) A novel imaging method for correlating 2D light microscopic data and 3D volume data based on block-face imaging. Sci Rep 7(1):3645
6. Ishii N, Tajika Y, Murakami T, Galipon J, Shirahata H, Mukai R, Uehara D, Kaneko R, Yamazaki Y, Yoshimoto Y, Iwasaki H (2021) Correlative microscopy and block-face imaging (CoMBI) method for both paraffin-embedded and frozen specimens. Sci Rep 11(1):13108
7. Tajika Y, Ishii N, Morimura Y, Fukuda K, Shikada M, Murakami T, Ichinose S, Yoshimoto Y, Iwasaki H (2023) Correlative microscopy and block-face imaging (CoMBI): a 3D imaging method with wide applicability in the field of biological science. Anat Sci Int 98(3):353–359

Chapter 4

Fetal Mouse Heart Clearing, Imaging, and 3D Reconstruction

Hamna Ammar and Paul Delgado-Olguín

Abstract

Traditional histological sectioning has been widely used to examine congenital heart defects (CHDs) in mouse models, providing detailed insights into tissue and cellular architecture. However, this approach is time-consuming, labor-intensive, and can miss subtle defects due to the critical need for precise serial sections. In contrast, three-dimensional imaging of cleared embryonic hearts using light sheet fluorescence microscopy offers a more efficient and comprehensive solution. This technique enables the rapid acquisition of z-stacks, including thousands of sections, without manual sectioning, which can be reconstructed into high-resolution 3D models. As a result, it significantly improves the detection and visualization of CHDs in mouse fetuses.

Keywords Tissue clearing, Congenital heart defects, Embryonic heart, 3D reconstruction, Light sheet microscopy

1 Introduction

Tissue clearing methodologies, in combination with light sheet fluorescence microscopy (LSFM), have significantly advanced three-dimensional (3D) imaging and streamlined visualization of large, complex tissues [1]. LSFM utilizes thin planes of light to optically section cleared or transparent tissues at a high resolution, allowing for the rapid acquisition of volumetric data that can be reconstructed into high-quality 3D models for in-depth analysis [2].

Although more commonly utilized in neuroscientific research, LSFM offers significant advantages for studying embryonic development, particularly in imaging the mouse fetal heart. Its ability to generate high-resolution, three-dimensional reconstructions of optically cleared tissues facilitates detailed visualization of complex cardiac structures, such as the outflow tract. Outflow tract defects are a common and clinically significant subset of congenital defects that can be challenging to classify with traditional histological methods alone [3]. Additionally, small ventricular septal defects at embryonic stages can be easily overlooked. LSFM addresses these limitations by

Paul Delgado-Olguín (ed.), *Mouse Embryogenesis*, Methods in Molecular Biology, vol. 3010,
https://doi.org/10.1007/978-1-0716-5126-1_4,

enabling more comprehensive and accurate detection and analysis of congenital heart defects. By capturing thousands of sections with z-step sizes typically ranging from 0.1 to 5 μm, LSFM provides the fine resolution necessary for detailed 3D reconstruction of cardiac structures, enabling the identification of subtle defects that may be missed with conventional histology [4, 5]. Here, we describe a protocol for the resection and dissection of mouse embryonic hearts for whole-heart imaging using a stereomicroscope. Furthermore, we detail an optimized ethyl cinnamate-based tissue clearing protocol compatible with LSFM. This approach allows for the imaging of fetal hearts using tissue autofluorescence, providing a clear and efficient means of visualizing cardiac structures without the need for external fluorescence markers.

2 Materials

2.1 Sample Preparation

Prepare all solutions using double-distilled (ddH_2O) or ultrapure water.

1. 1× Phosphate-Buffered Saline (PBS): Prepare 10× PBS by dissolving 160 g of NaCl, 4 g of KCl, 28.8 g of Na_2HPO_4, and 4.8 g of KH_2PO_4 in 1.8 L of water. Adjust pH to 7.4 by adding 5 M NaOH solution. Add additional water up to 2 L and autoclave to sterilize. Store at room temperature. Prepare 1× PBS by adding 50 mL of 10× PBS to 450 mL of water. Autoclave to sterilize. Store at room temperature.
2. 4% Paraformaldehyde (PFA): Prepare 20% PFA by adding 200 g of paraformaldehyde to 850 mL of water and 3 mL of 10 N NaOH. Heat the solution in a water bath at 65°C until clear. Cool down at room temperature and add water up to 1 L. Filter solution using sterile bottle-top filters. Aliquot into 50 mL conical tubes and store at −20°C (*see* **Note 1**).
3. 0.05% Sodium Azide (NaN_3) in PBS: Weigh 0.05 g of NaN_3 on an analytical scale and transfer it into a bottle containing 100 mL of autoclaved 1× PBS. Stir the solution on a magnetic stirring plate to ensure that NaN_3 is properly dissolved. The laboratory scale should be moved to the fume hood before weighing NaN_3, and the solution should be prepared in a well-ventilated area (*see* **Note 2**).
4. Heparin-KCl PBS: Prepare a 50 mL Heparin Solution at a final concentration of 1000 units/mL using the specific activity provided by the supplier. To prepare this solution of Heparin-KCl PBS, start by adding 10 mL of the 1000 units/mL heparin stock solution into 70 mL of 1× PBS. Add an additional 0.132 g of KCl to the solution and mix until dissolved. Add another 20 mL of 1× PBS in the solution to achieve a final

volume of 100 mL. Since 1× PBS already contains 2.68 mM KCl, adding an additional 1.32 mM of KCl with achieve a final concentration of 4 mM. Store at 4°C for future use.

5. Dissection microscope.
6. 90 mm petri dishes.
7. 90 mm Glass petri dish with black Sylgard coating.
8. 12-well plates.
9. 5 mL Disposable Transfer Pipettes.
10. 5 mL round-bottom tubes.
11. Tweezers.
12. Dissecting scissors.

2.2 Delipidation

1. B1N buffer: Prepare a 20% NaN_3 solution by dissolving 10 g of NaN_3 in 50 mL of ddH_2O. Use a magnetic stirrer to ensure the powder is well dissolved. To prepare the B1N buffer, dissolve 22.6 g of glycine in 1000 mL of ddH_2O to achieve a 0.3 M glycine solution. To this solution, add 1 mL of Triton X-100 and 0.5 mL of the 20% NaN_3 solution to achieve a final concentration of 0.1% Triton X-100 (v/v) and 0.01% NaN_3 (w/v). Mix and adjust pH to between 7 and 7.4 with NaOH (*see* **Note 2**).
2. 100% Methanol (*see* **Note 3**).
3. Dichloromethane (DCM) (*see* **Note 4**).
4. 5% Hydrogen peroxide (H_2O_2) in methanol: To make a 30 mL solution, dilute 5 mL of 30% H_2O_2 in 25 mL of 100% Methanol (*see* **Note 3**).
5. Dehydration Buffers: Prepare 50 mL solutions of Dehydration buffers 1–4 as indicated in Table 1 (*see* **Note 5**).

2.3 Agarose Embedding

1. Weighing boats.
2. Agarose.
3. 1× PBS.

Table 1
Percentages of absolute Methanol and B1N buffer to be used to prepare Dehydration buffers

Dehydration Buffer	100% Methanol	B1N Buffer
1	20%	80%
2	40%	60%
3	60%	40%
4	80%	20%

4. Dissection microscope.
5. Tweezers.

2.4 Tissue Clearing

1. 25% Methanol in ddH_2O.
2. 50% Methanol in ddH_2O.
3. 75% Methanol in ddH_2O.
4. Orbital shaker.
5. Tweezers.
6. Ethyl cinnamate.

2.5 Imaging

1. 90 mm Glass petri dish with black Sylgard coating.
2. Stereomicroscope for brightfield imaging (e.g., Nikon SMZ1000 Stereomicroscope™).
3. Light Sheet Fluorescence Microscope compatible with ethyl cinnamate (e.g., Miltenyi Biotec UltraMicroscope Blaze™).

3 Methods

3.1 Whole-Heart Imaging

Sample Preparation

1. Animal procedures must follow Institutional guidelines and be approved by the corresponding Animal Care Committee. Set up timed matings of mice and euthanize pregnant females at the desired stage. Ensure the clinical death of the mouse before harvesting the litter.
2. Harvest the uterine horns from the abdomen of a pregnant female using dissecting scissors and tweezers. Cut the uterine horns to dissect the embryo out of the uterus along with its yolk sac. Place it into ice-cold 1× PBS in a petri dish.
3. Carefully detach each embryo from its placenta and yolk sac using microdissection tweezers under a dissection microscope. Place embryos in a well of a 12-well plate with ice-cold 1× PBS. At this point, small tissue samples from the yolk sac or the embryo tail can be collected for genotyping.
4. Use the following steps to carefully dissect the heart out of the embryo, keeping the outflow tract preserved:
   (a) Using small pinching movements of microdissection tweezers, cut along the sternum and laterally along the rib cage, underneath the chest cavity, of the mouse embryo.
   (b) Peel away the ribcage and other tissue in the chest cavity to expose the heart.

(c) Sever any vessels at the bottom of the heart and remove the remaining membranes from the heart.

(d) Using the lungs, lift the heart from the back and pinch the top vessels. Gently lift the heart out of the chest cavity. Removing the heart attached to the trachea, and thymus will help preserve the vessels of the outflow tract.

(e) Cut the tip of a disposable transfer pipette and use it to transfer heart onto another dish.

(f) Carefully remove lungs, pericardial membrane, thymus, and trachea, being careful not to damage the vessels and atria on top of the heart.

(g) Transfer the dissected embryonic heart into a well of a 12-well plate with 1× PBS using a transfer pipette.

Embryos can be pinned down in a black Sylgard coating petri dish using stainless pins to ease dissections. Insert pins on both shoulders, hindlimbs, and on the head. Whole-heart images can also be taken within the chest cavity for embryos after pinning them down (ideal for all embryos younger than 12.5) (Fig. 1a, b).

5. Hearts can be placed in 4% PFA for long-term storage, further processed for H&E staining or other analyses.

Imaging

1. Create a hole in the dish big enough to fit the heart. Ensure that you remove any air bubbles that may form.
2. Remove heart from 1× PBS or 4% PFA and place in the hole of dish-oriented face-up.
3. Imagine the heart using a stereomicroscope and a light lamp for illumination.

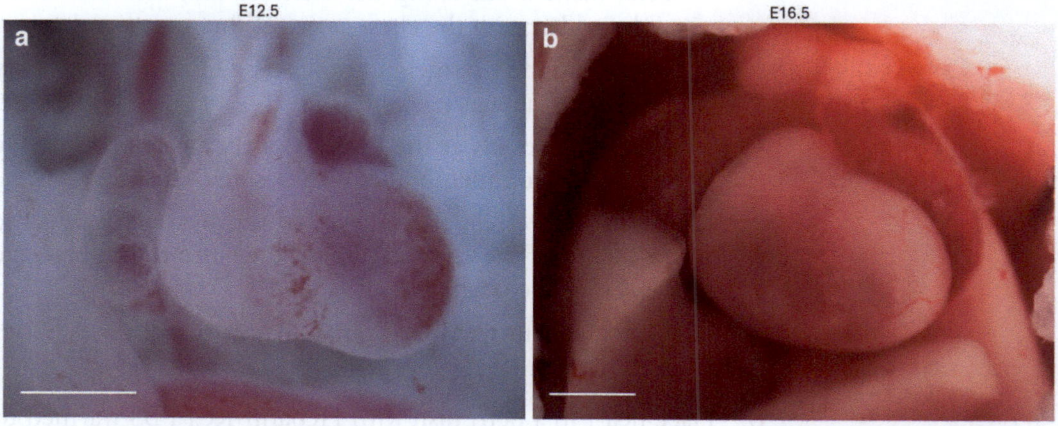

Fig. 1 Whole-heart images of mouse embryonic hearts. (**a**) Brightfield image of a mouse embryonic heart at E12.5 and (**b**) E16.5 taken within the chest cavity using the Nikon SMZ1000 Stereomicroscope. Scale bar = 1000 μm

4. Using an imaging software such as NIS-Elements, take two-dimensional flat or z-stack focused images of the whole heart. Adjust the illumination light and exposure to enable even lighting around the heart.

3.2 Tissue Clearing and 3D Reconstruction Using Light Sheet Microscopy

Sample Preparation

1. Pre-warm 100 units/mL heparin solution in a 37°C water bath and pre-cool 500 mL of 1× PBS 30 min prior to dissections.
2. Animal procedures must follow Institutional guidelines and be approved by the corresponding Animal Care Committee. Set up timed matings of mice and euthanize pregnant females at the desired stage. Ensure the clinical death of the mouse before harvesting the litter.
3. Harvest the uterine horns from the abdomen of a pregnant female using dissecting scissors and tweezers. Cut the uterine horns to dissect the embryo out of the uterus along with its yolk sac. Place it into ice-cold 1× PBS in a petri dish.
4. Carefully detach each embryo from its placenta and yolk sac using microdissection tweezers under a dissection microscope. Place embryos in a well of a 12-well plate with ice-cold 1× PBS. At this point, small tissue samples from the yolk sac or the embryo tail can be collected for genotyping.
5. Use the following steps to carefully dissect the heart out of embryo, keeping the outflow tract preserved:
   - (a) Using small pinching movements of microdissection tweezers, make a cut along the sternum and one lateral cut right underneath the chest cavity of the mouse embryo.
   - (b) Peel away the ribcage and other tissue in the chest cavity to expose the heart.
   - (c) Sever any vessels at the bottom of the heart and remove any remaining membranes on the side of heart.
   - (d) Pinch the vessels at the back of the heart and pull the heart out of the chest cavity, making sure to preserve the vessels of the outflow tract.
   - (e) Quickly remove lungs and major membranes from heart.
6. Place heart in a petri dish with Heparin-KCl PBS warmed to 37°C. Leave heart in solution for 10 min to allow full removal of blood (*see* **Note 6**).

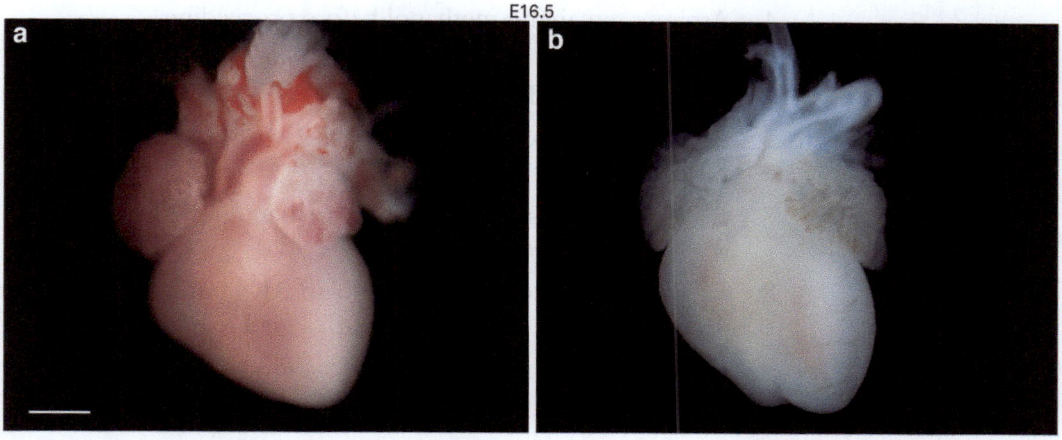

Fig. 2 Whole-heart images of resected mouse embryonic hearts at E16.5. (**a**) Brightfield image of a resected mouse embryonic heart attached to trachea and extracardiac membranes. (**b**) Brightfield image of a different resected mouse embryonic heart without attached trachea and extracardiac membranes. Images were taken using the Nikon SMZ1000 Stereomicroscope. Scale bar = 1000 μm

7. Immediately place the heart in ice-cold 1× PBS in a 90 mm petri dish. Carefully remove the remaining tissues near the outflow tract using dissection tweezers. Be careful not to damage the vessels and atria on top of the heart. Maintaining the integrity of vessels whilst removing the residual membranes is essential (Fig. 2a, b).
8. Transfer the dissected embryonic heart into a well of a 12-well plate with 4% PFA using a disposable pipette.
9. Repeat Steps 5–8 for all embryos.
10. Fix heart in 4% PFA overnight (ON) at 4°C inverting or shaking in a round-bottom tube.
11. Wash heart in 1× PBS three times (×) for 10 min each at 4°C.
12. Store hearts in 0.05% NaN_3 in PBS for long-term storage at 4°C (*see* **Note 7**).

Delipidation

All the following steps in this section should be carried out at 4°C with mild shaking. Use a 2 mL round-bottom tube with 1.6 mL of solution for incubation. Make sure the sample can freely move in the round-bottom tube during shaking. All buffers used in this section should be pre-chilled at 4°C before use.

1. Dehydrate the tissue using methanol in B1N buffer 20% > 40% > 60% > 80% at RT for 1 h each, shaking or rotating (*see* **Note 8**).
2. Incubate in 100% methanol for 1 h, 2× at 4°C.
3. Transfer all samples in a 5 mL glass vial using a transfer pipette.

4. Aspirate remaining methanol before proceeding.
5. Incubate sample in 2:1 DCM:100% methanol ON at 4°C (*see* **Notes 3, 4, and 9**).
6. Replace with DCM and incubate 2× for 30 min at 4°C (*see* **Note 10**).
7. Bleach sample using 5% H_2O_2 in methanol for 1 h or overnight at 4°C (*see* **Notes 11 and 12**).
8. Rehydrate sample using methanol in B1N buffer 80% > 60% > 40% > 20%. Incubate in each buffer for 1 h at 4°C.
9. Incubate the sample in B1N buffer for 30 min at room temperature (RT).

Agarose Embedding

1. Make 1% agarose by adding 0.5 g of agarose to a 50 mL 1× PBS solution. Heat solution until homogenized and allow solution to cool to 40°C (*see* **Note 13**).
2. Layer agarose solution in a small weight boat and position the sample in one corner, avoiding excess carryover of PBS. Gently pour another layer of agarose on top and, using tweezers, position the heart in the desired orientation (*see* **Notes 14 and 15**).
3. Loosen the agarose block from the weigh boat after it has solidified.
4. Transfer the sample to a clean platform and trim the agarose so that it forms the shape of a cube. Make sure to trim as close to the sample as possible. Transfer the sample back into the glass vial for further processing (*see* **Note 16**).

Tissue Clearing

The following steps should be carried out at room temperature with mild shaking.

1. Incubate sample in a gradient of methanol in water: 25% > 50% > 75% for 1 h each at RT.
2. Incubate sample in 100% methanol for 1 h each at RT 2×.
3. Incubate sample in 2:1 DCM:100% Methanol ON at RT.
4. Incubate the sample in DCM 3×, for 3 h each time.
5. Incubate the sample in ethyl cinnamate 3×, for 2 h each time.
6. Hearts should be left in ethyl cinnamate until tissue clears (*see* **Note 17**).

Imaging

1. Imaging with an ethyl cinnamate-compatible light-sheet microscope
   (a) Mount the agarose-embedded sample onto the sample holder using a silicone sealant or another adhesive (*see* **Note 18**).

(b) Immerse the sample in the ethyl cinnamate-containing imaging chamber and image using a dipping objective approved and matched to the refractive index of ethyl cinnamate for ethyl-cinnamate-immersed imaging (*see* **Note 19**).

(c) Image with a 1–2× magnification objective using multiple channels (Red: 590 nm, Green: 488 nm, Blue: 405 nm) (*see* **Notes 19 and 20**) (Fig. 3a–c).

3D Reconstruction Using Imaris

1. Ensure all z-stacks acquired from one experiment are organized into one file folder. TIFF images are recommended.
2. Image import using Imaris File Converter.
   (a) Open Imaris FileConverter.
   (b) Add files > select one image.
   (c) Select input > Start All.
3. 3D Reconstruction in Imaris
   (a) Select the .ims file created in the same folder as the TIFF stacks > open using Imaris.
   (b) Choose "3D Volume" to visualize the entire stack as a 3D object, adjusting transparency, channel thresholds, and coloring as needed.
   (c) Choose "Normal Shading" > Increase rendering quality as needed (Fig. 4a–c).
   (d) Use the "Slice" function to look at individual z-stack slices in the software (Fig. 3a–c).

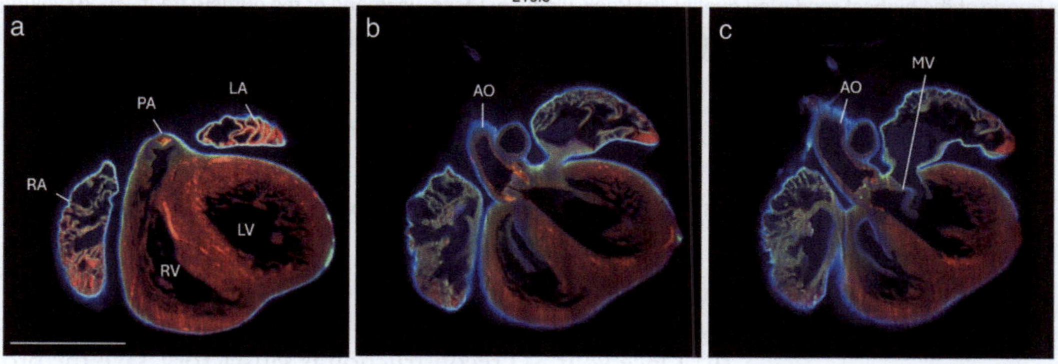

Fig. 3 Sagital z-stack sections of an ethyl cinnamate-cleared mouse embryonic heart at E16.5. Fluorescent images of a mouse embryonic heart at the level of the (**a**) pulmonary artery, (**b**) aorta, and (**c**) mitral valve. Serial z-stacks (3.5 μm) were obtained using a Blaze Light Sheet UltraMicroscope. Red, green, and blue signals were captured. AO = aorta, LA = left atrium, LV = left ventricle, MV = mitral valve, PA = pulmonary artery, RA = right atrium, RV = right ventricle. Scale bar = 1000 μm

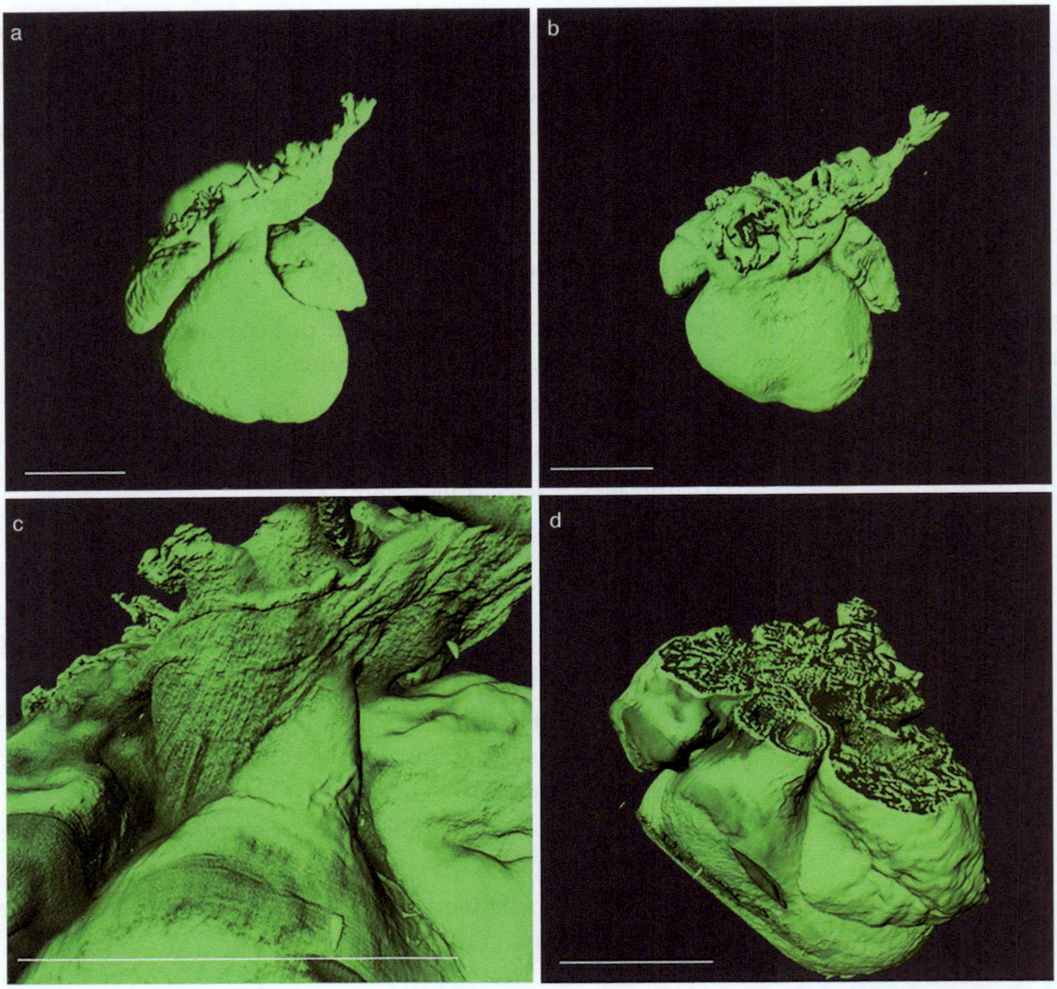

Fig. 4 Three-dimensional reconstructions of ethyl cinnamate-cleared mouse embryonic heart at E16.5. (**a**) High-resolution 3D volume rendering of the E16.5 mouse embryonic heart using the "Normal Shading" function in Imaris at the highest rendering quality. (**b**) Back-view. (**c**) Close-up image of the outflow tract. (**d**) 3D volume sectioned at the level of the outflow tract opening using the "clipping" tool. Scale bar = 1000 μm

(e) Use the "Snapshot" function to take images of the 3D volume (Fig. 4a–d).

(f) Use the "Clipping" function to section the 3D volume at the desired level (Fig. 4d).

(g) Create animations and videos using the "Animation" function.

(h) Once your 3D reconstruction is satisfactory, save the project as an Imaris file (.ims) for future editing and reference.

4 Notes

1. Paraformaldehyde is a very toxic and harmful irritant for the eyes, skin, and the respiratory tract. Wear proper protective equipment and work under a fume hood. Ensure that the PFA powder dissolves completely and allow the solution to cool down before filtration.
2. Sodium azide is toxic and should be handled with care. Always wear appropriate protective equipment and always work in a fume hood. Avoid direct contact with sodium azide and don't breathe in the dust. Sodium azide solution should be stored safely in a tightly sealed container to prevent accidental exposure and degradation.
3. Methanol is volatile, highly flammable, and toxic. Avoid inhalation and skin or eye contact.
4. Dichloromethane is highly volatile and should be handled under a fume hood. Only use glass containers for dichloromethane storage. Glass petri dishes and serological pipets must be used, as equipment made from polystyrene is not compatible with dichloromethane.
5. Glycine crystals may form in Dehydration buffers 3 and 4 due to saturation. Use the liquid solution for incubation and try to avoid drawing up the precipitated crystals.
6. Do not push on the heart. Allow the heart to stop contracting until most blood has been expelled. The heart can be kept on a heat pad to keep the solution warm.
7. Tissues can be stored in PBS with 0.05% NaN_3 at 4°C for several months to 1 year. Tissue quality may deteriorate following prolonged storage.
8. Tissues may appear to shrink slightly due to dehydration.
9. DCM is not compatible with polystyrene microcentrifuge tubes. DCM incubations should be done in glass scintillation vials or in polypropylene microcentrifuge tubes.
10. Tissues should sink to the bottom of the tubes at the end of each DCM wash to ensure full delipidation. Increase incubation time if necessary.
11. Hemoglobin from red blood cells can cause strong autofluorescence. If tissues are not well perfused, bleach with 5% H_2O_2 in methanol to reduce background.
12. Once DCM is removed at the end of each incubation, quickly apply fresh buffer from the next wash to avoid sample desiccation.

13. While the agarose is cooling down, gently swirl the solution so that it remains homogenous. Avoid exposing samples to excessively hot solutions.
14. The size of the weight boat or mold depends on tissue size and imaging chamber size.
15. Agarose embedding helps to stabilize the sample shape during imaging and prevents damage to the sample during imaging with a light sheet microscope.
16. Use a sharp single razor blade or a scalpel to trim agarose.
17. Samples should appear transparent in visible light at the end of clearing. After clearing, samples should be stored in the dark at room temperature until imaging.
18. Silicone glue is used to secure and mount specimens, providing a stable, transparent, and vibration-resistant bond for optimal imaging clarity.
19. Make sure to use a dipping objective compatible with ethyl cinnamate. Refer to the manufacturer's manuals for more detailed instructions. When imaging mouse embryonic hearts, the blue channel (405 nm excitation) highlights nuclei and connective tissue, the green channel (488 nm) detects collagen and elastin, and the red channel (560 nm) captures red blood cells, lipofuscins, and muscle tissue. Channel color may be altered later to enhance structures or to support 3D visualization [6].
20. Switch to an objective with higher magnification (e.g., 4–12× magnification) to zoom in on the regions of interest to image detailed structures.

Acknowledgments

This work was supported by the Heart and Stroke Foundation of Canada (G-17-0018613), the Natural Sciences and Engineering Research Council of Canada (NSERC) (500865), the Canadian Institutes of Health Research (CIHR) (PJT-149046), and Operational Funds from the Hospital for Sick Children to P.D.-O.

Conflict of Interest The authors have no conflicts of interest to declare that are relevant to the content of this chapter.

References

1. Gómez-Gaviro MV, Sanderson D, Ripoll J, Desco M (2020) Biomedical applications of tissue clearing and three-dimensional imaging in health and disease. iScience 23(8):101432
2. Swoger J, Pampaloni F, Stelzer EH (2014) Light-sheet-based fluorescence microscopy for three-dimensional imaging of biological samples. Cold Spring Harb Protoc 2014(1):1–8
3. Liu X, Kim AJ, Reynolds W, Wu Y, Lo CW (2017) Phenotyping cardiac and structural birth defects in fetal and newborn mice. Birth Defects Res 109(10):778–790
4. Ding Y, Bailey Z, Messerschmidt V, Nie J, Bryant R, Rugonyi S, Fei P, Lee J, Hsiai TK (2018) Light-sheet fluorescence microscopy for the study of the murine heart. J Vis Exp 139:57769
5. Fei P, Lee J, Packard RR, Sereti KI, Xu H, Ma J, Ding Y, Kang H, Chen H, Sung K, Kulkarni R, Ardehali R, Kuo CC, Xu X, Ho CM, Hsiai TK (2016) Cardiac light-sheet fluorescent microscopy for multi-scale and rapid imaging of architecture and function. Sci Rep 6:22489
6. Croce AC, Bottiroli G (2014) Autofluorescence spectroscopy and imaging: a tool for biomedical research and diagnosis. Eur J Histochem 58(4):2461

Chapter 5

Visualization of the Embryonic Great Vessels Using Plastic Resin Injection

Rimshah Abid, Yena Oh and Kyoung-Han Kim

Abstract

Accurate visualization of embryonic vasculature provides critical insights into cardiovascular development and the origins of congenital heart defects. Here, we describe a protocol for a low-viscosity plastic resin injection into the fetal mouse heart, which then polymerizes to create stable casts of the great vessels. These casts preserve fine vascular architecture and enable the detection of structural abnormalities, such as aortic arch artery anomalies and ventricular septal defects. This method provides a robust and reproducible approach for investigating both normal and pathological cardiovascular networks in rodent embryos.

Keywords Congenital heart disease, Arch artery defects, Plastic resin casting, Great vessels

1 Introduction

Congenital heart anomalies, such as ventricular septal defects (VSDs) and aortic arch artery malformations (e.g., coarctation and interrupted aortic arch), pose significant health risks and require detailed vascular visualization for understanding their origins and potential therapies [1–5].

Several methods exist for the visualization of vascular networks, each with its own advantages and limitations. Gelatin-Indian ink perfusion represents a commonly used technique for visualizing the cardiovascular and cerebral vasculature. Although this is a cost-effective method that provides visualization of vascular patterns, particularly for early embryos (e.g., E10.5), Indian ink has limitations, including premature solidification, the risk of vessel rupture, and poor reproducibility [6, 7]. Contrast agents serve as another method for visualizing vascular networks with high spatial resolution. Using long-circulating blood pool contrast agents, such as those based

The authors, Rimshah Abid and Yena Oh, contributed equally.

Paul Delgado-Olguín (ed.), *Mouse Embryogenesis*, Methods in Molecular Biology, vol. 3010,
https://doi.org/10.1007/978-1-0716-5126-1_5,

on iodine, gold, and bismuth, combined with X-ray-based imaging techniques, researchers can obtain uniform and stable visualization of vascular systems. However, these techniques are often costly and require specialized equipment and infrastructure [8–10].

In this chapter, we present a plastic resin casting method, which allows for the generation of detailed three-dimensional models of cardiovascular networks. This method, adapted from Liu and Martin [11], involves the precise injection of a low-viscosity plastic resin solution into the targeted vascular system, allowing it to polymerize and solidify within the vasculature. This process enables clear visualizaction of major arch arteries. Optionally, upon polymerization, surrounding biological tissue can be chemically macerated and dissolved, leaving behind an accurate and detailed cast of the vascular network. Utilizing this technique, we demonstrate its utility by examining arch artery malformations and ventricular septal defects in a genetic mouse model. This approach provides a powerful tool for studying normal vascular architecture, identifying vascular anomalies, and analyzing disease phenotypes in a robust three-dimensional manner.

2 Materials

2.1 Reagents

1. Batson's No. 17 Plastic Replica and Corrosion Kit. This kit contains Base Solution A, Catalyst, and Promoter C.
2. Phosphate-buffered saline (PBS) (1X, pH 7.4).
3. Water (H_2O) (sterile or distilled).
4. Maceration solution.
5. (Optional) 10% buffered formalin or 4% paraformaldehyde for fixation if samples are to be stored for future imaging.

2.2 Equipment and Supplies

1. Personal protective equipment, including gloves, safety goggles, and a lab coat.
2. Fume hood (mandatory for handling resin and promoter).
3. Stereomicroscope equipped with camera.
4. Fine dissection instruments (straight and curved forceps, small scissors, and a scalpel).
5. Micromanipulator (optional but recommended for fine control of injection).
6. Micropipette puller.
7. Glass capillary tubes (6-inch length, no filament).
8. Dissection dishes (silicone-coated dissection dishes).
9. Surgical tape (for securing the capillary pipette to the manipulator arm or pipette holder).
10. Tygon tubing.

11. Plastic transfer pipettes.
12. 1 mL and 10 mL syringes.
13. Gel-loading pipette tips (200 μL, thin tips for connecting the syringe to the capillary).
14. Incubator or water bath: preset to 50°C for tissue maceration.
15. (Optional) Bunsen burner or alcohol lamp, razor blade, and tweezers (for creating custom pipette tip fillers).

3 Methods

CAUTION: All steps involving the Batson's No. 17 Plastic Replica and Corrosion Kit must be carried out in a fume hood. Appropriate personal equipment, including gloves, safety goggles, and protective clothing, must be worn. Avoid direct skin contact and inhalation of fumes. Dispose of all waste materials according to institutional and environmental safety regulations.

3.1 Embryo Preparation for Resin Injection

1. Euthanize pregnant dam (e.g., E14.5 or E18.5) in accordance with institutional animal care and ethical guidelines (e.g., cervical dislocation).
2. Dissect the uterine horns: Expose the abdominal cavity, remove the uterine horns, and transfer them into a dish containing ice-cold PBS.
3. Isolate individual embryos: Under a stereomicroscope, gently remove extraembryonic membranes.
4. Arrest cardiac activity: Carefully sever the umbilical cord and place each embryo in ice-cold PBS to halt the heartbeat.
5. Expose the heart: Transfer one embryo at a time onto a silicone-coated dissection dish containing ice-cold PBS. Position the embryo ventral side up. Remove the sternum or ribs to expose the heart (*see* **Note 1**).

3.2 Plastic Resin Preparation and Capillary Pipette Loading

1. Under a fume hood, label a small beaker or polypropylene container. Ensure all resin components (Base Solution A, Catalyst, Pigment, Promoter) are equilibrated at room temperature.
2. Prepare the plastic resin mixture: For a final volume of 10 mL of final mixture, combine 10 mL of Base Solution A with ~0.2 mL of Pigment (blue or red) and 1.5 mL of Catalyst (*see* **Note 2**).
3. Mix the components thoroughly using a spatula until the solution is homogeneous.
4. Add 10–20 μL of Promoter C and mix slowly. The resin mixture will begin to polymerize within approximately 30–45 min (*see* **Note 3**).

5. Load the syringe: Draw ~1 mL of the plastic resin mixture into a 1 mL syringe. Avoid trapping air bubbles by pulling the plunger slowly.
6. Load the capillary: Attach a gel loading pipette tip to the syringe and fill a glass capillary pipette with the resin mixture (*see* **Note 4**).
7. Connect the capillary pipette to a Tygon tube that is connected to a 10 mL syringe. Secure the glass capillary pipette to the micromanipulator arm using tape (*see* **Note 5**).

3.3 Plastic Resin Injection into the Embryonic Heart

1. Place the embryo in a dissection dish and position it under a stereomicroscope. Keep the embryo submerged in ice-cold PBS (*see* **Note 6**).
2. Adjust the micromanipulator so that the capillary tip is aligned to the apex of the left ventricle.
3. Inject the plastic resin: Slowly and steadily push the syringe plunger, while observing the injection under the microscope as the resin fills the left ventricle and major arterial outflow tracts (*see* **Note 7**).
4. Hold the capillary in place for ~1 min to minimize backflow of resin, then gently withdraw the capillary.
5. Keep injected embryos in ice-cold PBS (*see* **Note 8**).
6. Repeat the procedure for remaining embryos, then image them using a stereomicroscope.

3.4 Optional: Tissue Corrosion for Stable Plastic Cast Generation

1. Initial curing: Place each injected embryo in ice-cold PBS or water for 30–60 min. Then, transfer the embryos to fresh water and incubate overnight at 4°C to allow complete polymerization of the injected resin.
2. Corrosion: Transfer embryos to a 6- or 12-well plate. Fully immerse each embryo in maceration solution.
3. Incubate at 50°C for 12 h without agitation.
4. Inspect: Rinse casts gently with water, replace with fresh Maceration solution, and continue incubation up to an additional 60 hours (72 h total) until all surrounding tissues are dissolved.
5. Rinse the cast with water to remove any remaining debris.
6. Document by photographing the final casts. Casts may be stored in water at 4°C for at least 1 year without noticeable degradation.

4 Notes

1. If genotyping is required, collect a small piece of tissue or tail from each embryo.

2. The described volume of plastic resin mixture is sufficient to inject approximately one to two mouse litters (i.e., ~20 E14.5 embryos) in a single session.

3. Adjust the amount of Catalyst or Promoter C to modify the polymerization rate.

4. Use a micropipette puller to generate a capillary glass from 6-inch, no-filament borosilicate glass tubing. A pulled capillary with a fine tip allows for precise injection into the embryonic heart.

5. A micromanipulator with a capillary pipette under a stereomicroscope is recommended for controlled injection. If this setup is unavailable, an insulin syringe (0.5 mL or 1.0 mL) can be substituted. In such cases, it is best to load plastic resin mixture directly into the syringe after removing the plunger to minimize bubbles.

6. Maintaining embryos in ice-cold PBS helps dissipate heat generated by the exothermic reaction of polymerization, thereby preventing potential thermal damage or distortion of the vascular cast.

7. Because the ductus arteriosus remains patent through mid-late gestation, injection into the left ventricle typically fills both systemic and pulmonary vasculature. If resin flows into the right

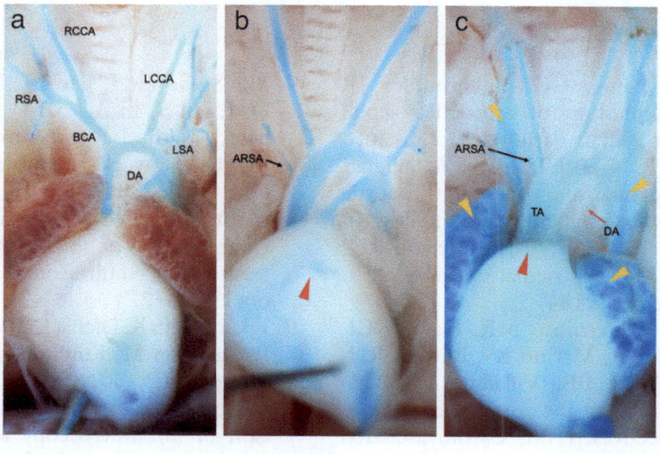

Fig. 1 Visualization of the arch arteries via plastic resin injections in E14.5 mouse embryos. (**a**) Representative embryo with a normal arch artery structure. Note that the resin remains confined to the left ventricle, indicating proper septation and intact valvular function. DA, ductus arteriosus; BCA, brachiocephalic artery; RSA, right subclavian artery; LSA, left subclavian artery; RCCA, right common carotid artery; LCCA, left common carotid artery. (**b**) A mutant embryo with mild congenital heart defects, displaying an aberrant RSA (ARSA; black arrow) and a ventricular septal defect (red arrowhead). (**c**) A mutant embryo with severe congenital heart defects, featuring ARSA (black arrow), truncus arteriosus (TA; red arrowhead), and a hypoplastic DA (red arrow). Due to an atrioventricular septal defect, the plastic resin flows retrogradely into the atria and superior vena cava (yellow arrowheads).

ventricle at E14.5 or later, this may indicate a ventricular septal defect or incomplete closure of the pulmonary valve to the ductus arteriosus (Fig. 1).

8. If immediate imaging is not feasible, fix the injected embryo in 10% buffered formalin or 4% paraformaldehyde. This preserves soft tissues prior to optional maceration or photographic documentation.

Acknowledgments

This work was supported by the Natural Sciences and Engineering Research Council of Canada (NSERC) Discovery Grant (RGPIN-2025-05100) and partly by the Canadian Institutes of Health Research (CIHR) Project Grant (PJT 173281 and PJT 195676) to Dr. Kyoung-Han Kim. Rimshah Abid was supported by the University of Ottawa Heart Institute Endowed Graduate Scholarship. Dr. Yena Oh was supported by the Frederick Banting and Charles Best Canada Graduate Scholarships Doctoral Award (CGS-D).

Conflict of Interest The authors have no conflicts of interest to declare that are relevant to the content of this chapter.

References

1. Anderson RH, Lenox CC, Zuberbuhler JR (1984) The morphology of ventricular septal defects. Perspect Pediatr Pathol 8(3):235–268
2. Priya S, Thomas R, Nagpal P, Sharma A, Steigner M (2018) Congenital anomalies of the aortic arch. Cardiovasc Diagn Ther 8(Suppl 1):S26–S44
3. Spicer DE, Hsu HH, Co-Vu J, Anderson RH, Fricker FJ (2014) Ventricular septal defect. Orphanet J Rare Dis 9:144
4. Bhatt AB, Lantin-Hermoso MR, Daniels CJ, Jaquiss R, Landis BJ, Marino BS, Rathod RH, Vincent RN, Keller BB, Villafane J (2022) Isolated Coarctation of the aorta: current concepts and perspectives. Front Cardiovasc Med 9:817866
5. Bae SB, Kang EJ, Choo KS, Lee J, Kim SH, Lim KJ, Kwon H (2022) Aortic arch variants and anomalies: embryology, imaging findings, and clinical considerations. J Cardiovasc Imaging 30(4):231–262
6. Aum DJ, Vellimana AK, Singh I, Milner E, Nelson JW, Han BH, Zipfel GJ (2017) A novel fluorescent imaging technique for assessment of cerebral vasospasm after experimental subarachnoid hemorrhage. Sci Rep 7(1):9126
7. Xue S, Gong H, Jiang T, Luo W, Meng Y, Liu Q, Chen S, Li A (2014) Indian-ink perfusion based method for reconstructing continuous vascular networks in whole mouse brain. PLoS One 9(1):e88067
8. Mehta KS, Lee JJ, Taha AG, Avgerinos E, Chaer RA (2017) Vascular applications of contrast-enhanced ultrasound imaging. J Vasc Surg 66(1):266–274
9. Epah J, Pálfi K, Dienst FL, Malacarne PF, Bremer R, Salamon M, Kumar S, Jo H, Schürmann C, Brandes RP (2018) 3D Imaging and quantitative analysis of vascular networks: a comparison of ultramicroscopy and micro-computed tomography. Theranostics 8(8):2117–2133
10. Starosolski Z, Villamizar CA, Rendon D, Paldino MJ, Milewicz DM, Ghaghada KB, Annapragada AV (2015) Ultra high-resolution in vivo computed tomography imaging of mouse cerebrovasculature using a long circulating blood pool contrast agent. Sci Rep 5:10178
11. Liu C, Martin J (2010) Visualizing fetal mouse great blood vessels by plastic casting. Cold Spring Harb Protoc 2010(5):pdb.prot5433

Chapter 6

Explant Culture for Studying Lung Development

Behzad Yeganeh and Martin Post

Abstract

Lung development is a complex process that requires the input of various signaling pathways to coordinate the specification and differentiation of multiple cell types. Ex vivo culture of the lung is a highly useful technique that represents an attractive model for investigating various processes critical to lung development, function, and disease pathology. Ex vivo cultured lungs remain comparable to the in vivo lung both in structure and function, which makes them more suitable than cell cultures for physiological studies. Lung explant cultures offer several significant advantages for studying morphogenic events that guide lung development, including budding, branching, and fusion. Additionally, these cultures preserve the native physiological interactions between cells in the developing lung, enabling investigations into the direct and indirect signalling taking place between tissues and cells throughout the developmental process. Studying the temporal and spatial control of gene expression by transcriptional factors using different reporters to understand their regulatory function at different moments of development is another valuable advantage of lung explant culture.

Keywords Lung branching morphogenesis, Organ culture, Air–liquid interface, Lung slice, Ex vivo culture

1 Introduction

The idea of culturing tissue or an organ outside of the body, a technique called explant culture, has its origin in the early twentieth century [1, 2]. The first studies using ex vivo cultures have shown that embryological tissues differentiate and behave like what appears to be normal tissue, indicating that we can use this technique to investigate fetal and embryological development [3].

Explant culture is a valuable technique for studying various processes implicated in organogenesis. An eminent advantage of this technique lies in the preservation of histotypic architecture and native physiological interactions among cells within the tissue, thereby facilitating observation of direct and indirect signaling events occurring between tissues and cells throughout developmental phases [4]. Additionally, explant culture facilitates the characterization of

Paul Delgado-Olguín (ed.), *Mouse Embryogenesis*, Methods in Molecular Biology, vol. 3010,
https://doi.org/10.1007/978-1-0716-5126-1_6,

morphogenic events crucial for embryonic organ development, including budding, branching, and fusion. At the cellular level, the explant technique allows for a closer examination and manipulation of cell migration, proliferation, and differentiation dynamics of specific cells [5, 6]. Ex vivo culture of explants is also useful to monitor gene expression patterns using various reporters, and understand their regulatory functions, as well as identifying the secretion of signaling molecules at different developmental stages [5–7].

Due to their size, the entire organ of an embryo can be used as an explant. It is suggested that such explant is more resistant to hypoxic conditions than mature tissue. Therefore, the ex vivo explant culture technique has predominantly been used for developmental and embryogenic studies [8]. However, in 1959, Trowell developed a protocol for the use of mature organs [3, 8]. This method is valuable for investigating pharmacological response variations in order to develop new therapies. And, it has been used to observe tissue reactions in the presence of infection [9].

The explant culture technique is not without limitations. The organ alone does not fully replicate the physiological environment complexity of a whole organism, and they are not suitable for long-term experiments. In those cases, organ transplantation is a more suitable alternative. Nevertheless, the ex vivo culturing technique offers significant advantages: researchers gain precise control over the media composition, manipulate both the tissue and the environment to create the perfect conditions to observe the effects of diverse growth factors and molecules under reproducible conditions [5]. This precise control increases the reliability and reproducibility of the results and contributes to a deeper understanding of cellular and molecular processes in organ development and function.

The explant culture technique has been applied to multiple organs, including the lung. Embryonic lung explants adopt a flattened morphology that enables the quantification of terminal branches and branch length. This technique has helped the study of some pivotal regulators involved in the budding and branching of the lung, such as fibroblast growth factor (FGF), bone morphogenetic protein (BMP), and the wingless type (Wnt) signaling family [10], as well as cell death pathways [11]. Also, it has been instrumental in elucidating the intricate epithelial and mesenchymal interactions that regulate lung embryogenesis, allowing for a better characterization of growth factors and signaling molecules involved in these interactions [10]. Ex vivo explants could be established using the entire lung or only the mesenchymal and epithelial components in order to further characterize their functions [10]. The explant culture technique has been and will continue to be an invaluable tool to further our knowledge on the cellular and molecular mechanisms controlling lung development.

2 Materials

2.1 Embryonic (E11.5) Mouse Lung Explant Culture

1. Dulbecco's Modified Eagle Medium: Nutrient Mixture F-12 (DMEM/F-12).
2. Hank's Buffered Salt Solution ($HBSS^-$, no calcium, no magnesium).
3. Penicillin/streptomycin.
4. L-Ascorbic acid.
5. Nuclepore™ Track-Etched Polycarbonate Membrane Filter (8 μm pore size, Whatman).
6. Plastic disposable transfer pipettes, sterile.
7. Flat-bottom cell culture plates with lids, 6-well plates.
8. Stereoscopic dissecting microscope.
9. Dissection tools, including Fine iris scissors (straight), Noyes Spring scissors, and Dumont forceps.
10. 50 mL conical centrifuge tubes.
11. Petri dishes with clear lid, 100 × 15 × 15 mm.
12. Insulin syringe with detachable needle.

2.2 E17.5 and Adult Mouse Lung Explant Culture

1. Low Melting Point Agarose.
2. Dulbecco's Modified Eagle Medium: Nutrient Mixture F-12 (DMEM/F-12).
3. Hank's Buffered Salt Solution (HBSS).
4. Dextrose (D-Glucose) anhydrous.
5. Penicillin/streptomycin.
6. L-Ascorbic acid.
7. Nucleopore hydrophobic floating membranes (8 μm pore size, Whatman).
8. Dissection tools, including Fine iris scissors (straight), Noyes Spring scissors, and Dumont forceps.
9. Flat-bottom cell culture plates with lids, 6-well plates.
10. Stereoscopic dissecting microscope.
11. Vibrating blade microtome.
12. 50 mL conical centrifuge tubes.
13. Petri dishes with a clear lid, 100 × 15 mm.

Animal experiments were carried out in accordance with the Animal Care Committee guidelines of the Hospital for Sick Children Research Institute.

3 Methods

Development of the lung in the mouse embryo is a complex process that begins on embryonic day 9.5 (E9.5), upon initiation of the respiratory diverticulum or lung buds from the ventral foregut endoderm and continues throughout the early postnatal stages. Based on histological morphology, mouse lung development and maturation have been divided into five stages: embryonic stage (E9–12), pseudoglandular (E12–14.5), canalicular (E14.5–16.5), terminal saccular (E16.5 to PD5), and alveolar or postnatal stage (PD5 to PD30). In order to determine the key molecules involved in the regulation of lung branching morphogenesis, lung organ culture can be performed at different stages of lung development. In the following section, we will describe two protocols used for culturing lung explants at early (E11.5) and late (E17.5) lung developmental stages.

3.1 Embryonic Lung Isolation and Culture

1. Set up timed mating with embryonic day zero (E0) defined as the day of vaginal plug. On postcoitum day 11.5 (E11.5), euthanize the timed-pregnant mice using CO_2 (*see* **Note 1**).
2. Under sterile conditions with appropriate skin disinfectant (*see* **Note 2**) (the abdomen area could be disinfected with 70% ethanol), remove the uterus from the animal in a laminar flow hood and place it in a 50 mL Falcon tube containing pre-cooled Hanks Balanced Salt Solution (HBSS) on ice.
3. Place the uterus in a Petri dish containing cold HBSS under the stereoscopic dissecting microscope and release the embryos from the uterus (*see* **Note 3**) (by incising the uterine wall using sterile spring scissors).
4. Transfer the embryos using a sterile squeezable Transfer Pipette in a new Petri dish containing cold HBSS and place it on ice (*see* **Note 4**) (cut the tip of the plastic Transfer Pipette according to the size of the embryos).
5. Under the dissecting microscope, using two 1 mL syringes with $29^{1/2}$ G needles, open the embryo along the spinal cord axis, slightly ventral to the spine, by making small tears with one needle while holding the body with the other needle.
6. Locate the primitive lungs in the ventral portion of the embryo, proximal to the incision, below the heart and above the stomach. With the needles, trim away the tissue to free the lungs (*see* **Note 5**) (embryonic lungs at day 11.5–12.5 in a mouse look like an inverted "Y" with asymmetric "arms" with buds starting to appear on the two bronchi).
7. Under the dissecting microscope, transfer the dissected lungs to a Petri dish containing cold HBSS using a sterile transfer pipette.

8. Prepare 6- or 12-well plates by filling the wells with 1 or 0.5 mL of DMEM/F-12, respectively (*see* **Note 6**) (supplemented with 50 units/mL of penicillin–streptomycin and 0.25 mg/mL ascorbic acid).
9. Place each dissected lung from HBSS onto a Nuclepore™ Track-Etched Polycarbonate Membrane Filter (8 μm pore size) using a sterile transfer pipette. Adjust the position of the dissected lungs on the membrane, spreading them evenly on the membrane with their trachea having a flattened and straight position.
10. Gently lay the membranes on the surface of the media wells, allowing the membranes to float on top of the media, creating an air-liquid culture setup.
11. Maintain the culture under optimal humidity with an atmosphere of 95% air and 5% CO_2 for the desired culture time.
12. Change the medium every day. Pictures can be taken at different time points. An example of the mouse embryonic lung explants culture and result produced is shown in Fig. 1.

3.2 E17.5 and Adult Mouse Lung Explant Culture

One of the major advantages of early lung embryonic explant culture at an air–liquid interface is that the tissue retains a spherical geometry when the medium is maintained at the correct level. However, due to the absence of the entire circulatory system, gaseous diffusion, exchange of nutrients and metabolites are limited to the periphery of the explant. The rate of diffusion limits the size of the explant

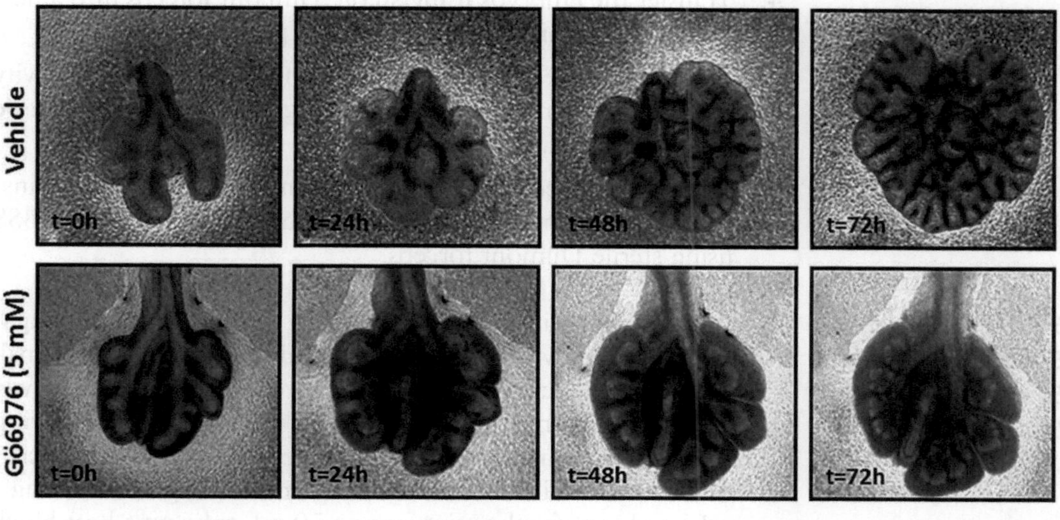

Fig. 1 Effects of Gö6976, a protein kinase C (PKC) inhibitor on in vitro branching morphogenesis of E11.5 lung. Lung explants were treated with Gö6976 (5 μM) or vehicle control for 72 h. Afterwards, the lungs were harvested to examine morphology and the number of terminal buds. Examination of the lung explants by light microscopy revealed less branches in the Gö6976-treated rudiments when compared to the control

culture. Thus, at later stages of development, the lung explant culture protocol requires some modifications. E17 or E18 lungs can be chopped into 0.5-mm cubes and cultured submerged in medium [12, 13]. Alternatively, these small pieces of the lungs (0.5-mm cubes) can be cultured on a permeable membrane at an air–liquid interface [14]. Bronchial epithelial cells of explants cultured in a submerged system fail to undergo mucociliary differentiation [15, 16], while small pieces of lungs used under air–liquid interface conditions are not representative of the whole structure of the lungs. To overcome these limitations, air–liquid interface cultures using a lung slice is another technique that provides a unique approach to investigate in vitro lung physio-pharmacotoxicology [17]. Living lung slices are robust and retain many aspects of the cellular and structural organization of the lung, including intrapulmonary airways, arterioles, and veins present within the alveolar parenchyma.

1. Euthanize timed-pregnant mice on day E17.5, using a CO_2 chamber, as described previously (*see* **Note 7**). (Clinical death of the animal must be ensured.)
2. Spray the abdomen area with 70% ethanol and remove the uterine horns containing embryos from the animal under sterile conditions in a laminar flow hood. Place the embryos in a 50 mL Falcon tube containing precooled Hanks Balanced Salt Solution (HBSS) on ice.
3. Place the uterus in a Petri dish containing cold HBSS under the stereoscopic dissecting microscope and release the embryos from the uterus by incising the uterine wall using sterile spring scissors.
4. Transfer the embryos using sterile Dumont forceps into a new Petri dish containing cold HBSS and place it on ice.
5. Remove the head of the embryo and open the chest cavity using Noyes spring scissors. The lung with three lobes can be easily visualized.
6. Under the dissecting microscope, remove the lungs and transfer them one by one to a Petri dish containing cold HBSS using sterile Dumont forceps.

3.3 Agarose Embedding (Prepare 30 min Before Dissections)

1. Prepare approximately 25 mL of 4% (w/v) low-melting agarose in HBSS with glucose for embedding all lobes into small rectangular blocks (*see* **Note 8**). To prepare the agarose: slowly add 1 g of low-melting agarose to 25 mL HBSS, while stirring rapidly with a magnetic bar. Remove the stir bar and microwave until it boils to dissolve the agarose. Add glucose (5 mg/mL) and transfer the mixture to a 50 mL tube on a heat block and maintain the temperature above 40°C to avoid gelling.

2. Set up the vibratome and the dissection area by sterilizing blades, weighing spatulas, embedding molds, a scalpel, a fine paintbrush, small scissors, forceps, and a small ice bucket with a sterile Petri dish for collecting sections.
3. Working in ice-cold HBSS buffer, dissect the lungs at the end of each lobar bronchi to detach each lobe using small scissors.
4. Pat dry each lobe using absorbent sheets to remove excess HBSS. Place each lobe in small metal base molds with at least 3 mm of 4% (w/v) low-melting-point agarose surrounding the tissue from the edges. Orient each lobe using forceps by positioning the lobe's largest flat edge at the surface of the metal mold facing the experimenter (Fig. 2a) (*see* **Note 9**). This edge will be fixed to the specimen plate for vibratome sectioning. Place the embedding molds on ice to gel the agarose.
5. Allow the blocks to harden on ice for at least 5 min prior to sectioning with the vibratome.

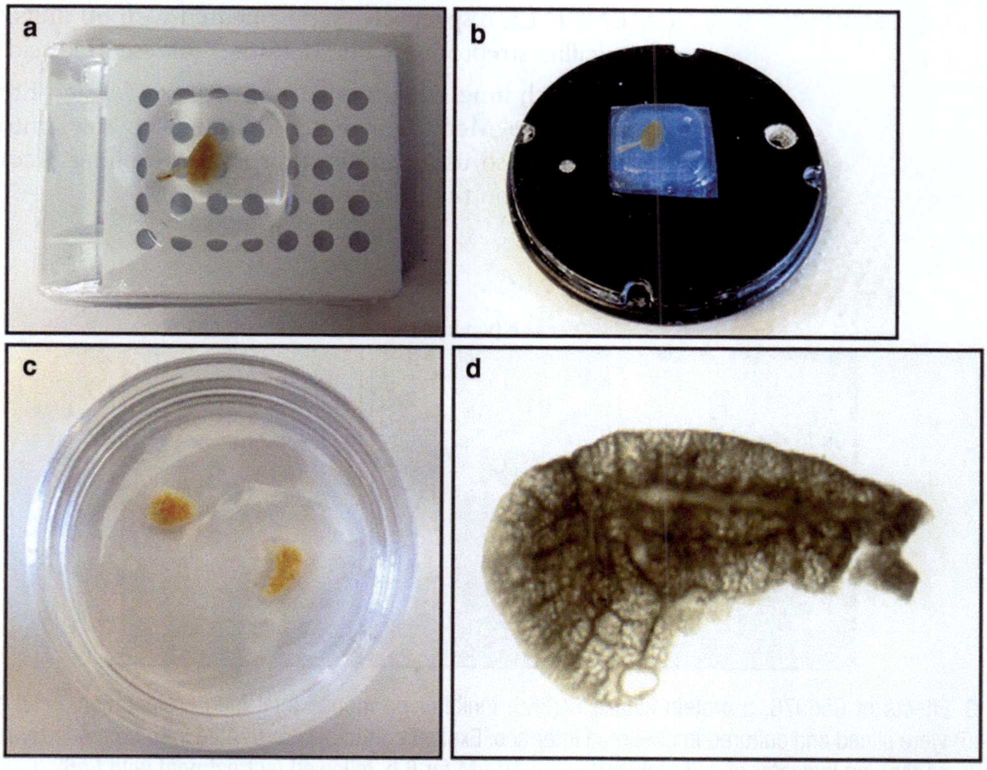

Fig. 2 Agarose embedding and lung slicing. (**a**) A single lobe of fetal lung is embedded in low-melting-point agarose in the mold. (**b**) Agarose block is removed from the mold and fixed onto the specimen mounting disk using adhesive mounting material. (**c**) Two lung slices collected in a Petri dish containing cold HBSS. (**d**) Representative lung slice. Lung slice under a stereoscopic dissecting microscope

3.4 Lung Slicing and Culture

1. Set up the vibratome by filling the sectioning chamber with ice-cold HBSS and maintain cold temperature throughout sectioning with the surrounding ice bath.
2. Remove the agarose block from the mold and trim down excess agarose surrounding the lobes, while keeping approximately 3 mm from the edge of the tissue.
3. Clean and dry the surface of the specimen mounting disk. Fix the tissue block onto the specimen mounting disk using sufficient adhesive mounting material and submerge the plate into the HBSS-filled sectioning chamber (Fig. 2b).
4. Set up sectioning on 350 μm and slice the agarose block to obtain 350 mm thick slices.
5. Collect sections with spatulas into DMEM/F-12 supplemented with 50 units/mL of penicillin–streptomycin in a sterile Petri dish on ice. Figure 2c shows an example of a lung slice collected in a Petri dish containing cold HBSS.
6. Manually remove the excess agarose from the sliced tissue using small scissors, and use the slice for air–liquid culture.
7. Prepare 6- or 12-well plates by filling wells with 1 or 0.5 mL of DMEM/F-12, respectively (supplemented with 50 units/mL of penicillin–streptomycin and 0.25 mg/mL ascorbic acid).
8. Transfer each lung section onto a Nuclepore™ Track-Etched Polycarbonate Membrane Filter (8 μm pore size) into an empty Petri dish using sterile forceps. Ensure lung slices are spread evenly on the membrane.

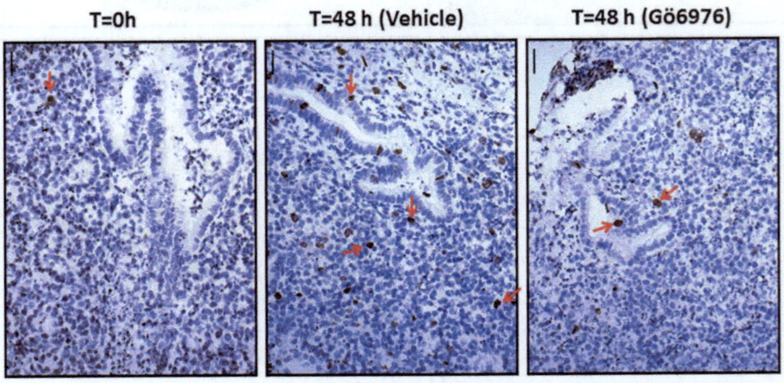

Fig. 3 Effects of Gö6976, a protein kinase C (PKC) inhibitor on mouse fetal lung proliferation. Fetal lungs (E17.5) were sliced and cultured in air–liquid interface. Explant cultures were treated with bromodeoxyuridine (BrdU) reagent, an indicator of cell proliferation, at 10 μM for 6 h, followed by treatment with Gö6976 (5 μM) or vehicle control for 48 h. Afterwards, the lung slices were harvested, fixed in 4% paraformaldehyde, and sectioned. Immunohistochemistry was carried out on sections with a monoclonal antibody specific for BrdU. Immunohistochemistry of the slices for BrdU revealed a reduced number of BrdU-positive cells (red arrows, dark brown nuclear staining) in the Gö6976-treated lung slices when compared to the control

9. Gently lay the membranes onto the surface of the media, allowing the membrane to float on top of the media, creating an air-liquid culture. Slices must be in the incubator within 2 h following initial dissection.
10. Maintain the culture under optimal humidity with an atmosphere of 95% air and 5% CO_2 for the desired culture time.
11. Change the medium every day. Slices can be maintained in vitro for 72 h, depending on the treatment. Pictures can be taken at different time points, or lung slices can be harvested and examined via histology. An example of the lung slice (E17.5) culture and result produced is shown in Fig. 3.

4 Notes

1. Place the animal in a chamber and expose the animal to 100% CO_2 at a fill rate of 10–30% of chamber volume per minute. Following unconsciousness, observe the animal for faded eye color and lack of respiration. When both are observed, maintain CO_2 filling for 1–2 min and then remove the animal from the chamber.
2. The abdominal area of the animal could be disinfected with 70% ethanol.
3. Use a sterile spring scissor to incise the uterine wall and release the embryos from the uterus. A timed-pregnant female yields an average of 8–10 embryos.
4. Cut the tip of a disposable plastic transfer pipette according to the size of the embryos. Transfer one embryo at a time by gently squeezing the plastic Transfer Pipette with your fingers.
5. Embryonic lungs at day 11.5–12.5 in a mouse look like an inverted "Y" with asymmetric "arms" with buds starting to appear on the two bronchi.
6. Prior to use, supplement the DMEM/F-12 medium with 50 units/mL penicillin–streptomycin and 0.25 mg/mL ascorbic acid.
7. Clinical death of the animal must be confirmed using an appropriate method, such as verification of cardiac and respiratory arrest or the presence of fixed, dilated pupils, followed by the application of a physical method to ensure death.
8. To prepare the agarose: slowly add 1 g of low-melting agarose to 25 mL HBSS, while stirring rapidly with a magnetic bar. Remove the stir bar and microwave until it boils to dissolve the agarose. Add glucose (5 mg/mL) and transfer the mixture to a 50 mL tube on a heat block and maintain the temperature above 40°C to avoid gelling.

9. This edge will be fixed to the specimen plate for vibratome sectioning.

Conflict of Interest

The authors have no conflicts of interest to declare that are relevant to the content of this chapter.

References

1. Harrison RG (1906) Observations of the living developing nerve fiber. Proc Soc Exp Biol Med 4(1):140–143
2. Carrel A (1912) On the permanent life of tissues outside of the organism. J Exp Med 15(5):516–528
3. Randall KJ, Turton J, Foster JR (2011) Explant culture of gastrointestinal tissue: a review of methods and applications. Cell Biol Toxicol 27(4):267–284
4. Narhi K, Thesleff I (2010) Explant culture of embryonic craniofacial tissues: analyzing effects of signaling molecules on gene expression. Methods Mol Biol 666:253–67
5. Narhi K (2017) Embryonic explant culture: studying effects of regulatory molecules on gene expression in craniofacial tissues. Methods Mol Biol 1537:367–80
6. Staneva R, Barbazan J, Simon A, Vignjevic DM, Krndija D (2018) Cell migration in tissues: explant culture and live imaging. Methods Mol Biol 1749:163–173
7. Sessions GA, Copp ME, Liu JY, Sinkler MA, D'Costa S, Diekman BO (2019) Controlled induction and targeted elimination of p16(INK4a)-expressing chondrocytes in cartilage explant culture. FASEB J 33(11):12364–12373
8. Trowell OA (1959) The culture of mature organs in a synthetic medium. Exp Cell Res 16(1):118–147
9. Nicholas B, Staples KJ, Moese S, Meldrum E, Ward J, Dennison P, Havelock T, Hinks TS, Amer K, Woo E, Chamberlain M, Singh N, North M, Pink S, Wilkinson TM, Djukanović R (2015) A novel lung explant model for the ex vivo study of efficacy and mechanisms of anti-influenza drugs. J Immunol 194(12):6144–6154
10. Del Moral PM, Warburton D (2010) Explant culture of mouse embryonic whole lung, isolated epithelium, or mesenchyme under chemically defined conditions as a system to evaluate the molecular mechanism of branching morphogenesis and cellular differentiation. Methods Mol Biol 633:71–79
11. Yeganeh B, Lee J, Ermini L, Lok I, Ackerley C, Post M (2019) Autophagy is required for lung development and morphogenesis. J Clin Invest 129(7):2904–19
12. Shan L, Emanuel RL, Dewald D, Torday JS, Asokanathan N, Wada K, Wada E, Sunday ME (2004) Bombesin-like peptide receptor gene expression, regulation, and function in fetal murine lung. Am J Physiol Lung Cell Mol Physiol 286(1):L165–L173
13. Sunday ME, Hua J, Dai HB, Nusrat A, Torday JS (1990) Bombesin increases fetal lung growth and maturation in utero and in organ culture. Am J Respir Cell Mol Biol 3(3):199–205
14. Prince LS, Okoh VO, Moninger TO, Matalon S (2004) Lipopolysaccharide increases alveolar type II cell number in fetal mouse lungs through Toll-like receptor 4 and NF-kappaB. Am J Physiol Lung Cell Mol Physiol 287(5):L999–L1006
15. Prytherch Z, Job C, Marshall H, Oreffo V, Foster M, BeruBe K (2011) Tissue-Specific stem cell differentiation in an in vitro airway model. Macromol Biosci 11(11):1467–77
16. Shojaie S, Ermini L, Ackerley C, Wang J, Chin S, Yeganeh B, Bilodeau M, Sambi M, Rogers I, Rossant J, Bear CE, Post M (2015) Acellular lung scaffolds direct differentiation of endoderm to functional airway epithelial cells: requirement of matrix-bound HS proteoglycans. Stem Cell Rep 4(3):419–430
17. Morin JP, Baste JM, Gay A, Crochemore C, Corbiere C, Monteil C (2013) Precision cut lung slices as an efficient tool for in vitro lung physio-pharmacotoxicology studies. Xenobiotica 43(1):63–72

Chapter 7

Embryonic Gut Explant Model for Enteric Neural Crest Cells Migration

Dorothy Lee and Agostino Pierro

Abstract

Migration of enteric neural crest cells (ENCCs) within the developing embryonic gut is key to the formation of the enteric nervous system (ENS). Investigating factors that guide the migration of these cells along the gut tube could aid in understanding the pathogenesis and developing potential therapy for ENS developmental defects, such as those present in Hirschsprung disease. However, limitations exist on appropriate study models to allow the live tracing of these migrating cells within the developing intestine. This protocol describes a mouse embryonic gut explant model utilizing a *Sox10*-Venus transgenic mouse to fluorescently label migrating ENCCs within the gut for live-cell imaging during culturing.

Keywords Enteric nervous system (ENS), Enteric neural crest cells (ENCCs), Cell migration, Explant, Live imaging

1 Introduction

The enteric nervous system (ENS) is a complex neural network that is composed of neurons and glial cells embedded between the muscle layers of the intestinal wall [1]. To understand the pathophysiology of ENS developmental defects such as Hirschsprung disease, various research models were established in the past, including in vivo genetically modified mouse models [2], in vitro cell culture-based models [3] or suspension cultured neurospheres [4]. A key progress driving the morphogenesis of the ENS is the migration of enteric neural crest cells (ENCCs) and neural progenitor. While in vivo animal models enable the assessment of survival and intestinal motility function, studying developmental processes in embryos during embryogenesis remains challenging. On the other hand, assessing migration using in vitro culture-based models requires the isolation of enteric neural progenitors and ENCCs, separating these target cells from the environment they react with and rely on to assess migration [5]. For this reason, in this chapter, we have established an embryonic gut

Paul Delgado-Olguín (ed.), *Mouse Embryogenesis*, Methods in Molecular Biology, vol. 3010,
https://doi.org/10.1007/978-1-0716-5126-1_7,

explant model to develop an intermediate model for live imaging to trace ENCCs migration while, at the same time, allowing these cells to stay within their microenvironment.

Visualization of the ENS during development is possible with the development of *Sox10-Venus* reporter mice [6], using *Sox10* as a neural crest cell marker and its expression to drive the expression of the fluorescence Venus protein. Upon dissection and isolation of the whole length embryonic gut from the *Sox10-Venus+* embryo, we use Matrigel to embed the embryonic gut in place within the cell culture dish. This allows the gut tissue to be fixed down and prevents it from floating after culture media is added. Advantages of using Matrigel includes its function as a gel matrix for 3-dimensional cell migration, as well as its similarity with the intestinal basement-membrane composition [7]. This explant model can be used as a screening model for ENCCs migration-inducing molecules through supplementation of the culture media with additional factors.

2 Materials

All reagents and materials should be prepared under sterile conditions to prevent contamination.

2.1 Animals

1. *Sox10-Venus* mice (*see* **Note 1**).

2.2 Equipment

1. Dissection stereo microscope with fluorescence (*see* **Note 2**).
2. Inverted epifluorescence microscope with cell culture plate holder and incubator chamber (*see* **Note 3**).
3. Fine microdissection forceps (Tips dimension: 0.05 × 0.02 mm) (*see* **Note 4**).
4. Heating pad.
5. Biosafety cabinet (BSC).
6. Tissue culture incubator at 37°C supplied with 5% CO_2.

2.3 Tubes and Culture Plates

1. 6 cm petri dishes.
2. Low-retention wide-bore P1000 pipette tips.
3. Low-retention P200 pipette tips.
4. 96-well cell culture plate.
5. 12-well cell culture plate.

2.4 Reagents

1. Phosphate-buffered saline (PBS).
2. Matrigel growth-factor reduced (*see* **Note 5**).
3. Explant culturing media: Dulbecco's modified Eagle's medium/Nutrient Mixture F-12 (DMEM/F-12), 1% penicillin/streptomycin (*see* **Note 6**).
4. Migration-inducing factor, chemoattractant (*see* **Note 7**).

3 Methods

3.1 Embryonic Gut Tissue Dissection

1. Following institutional guidelines and Animal Care Committee approval, euthanize pregnant dam at embryonic day (E) 10.5–14.5 and immediately extract the uterus and place it in a 6 cm petri dish with cold PBS (*see* **Note 8**).
2. Extract whole embryos from the uterus under a dissecting stereo microscope with fluorescence. To free the whole embryo, carefully cut the umbilical cord as close to the placenta as possible without disrupting the embryonic gut (*see* **Note 9**). Remove the yolk sac and amnion sac. Select embryos with *Sox10-Venus* positive expression for the explant ENCCs migration experiment.
3. Transfer embryos to a new 6 cm petri dish with cold PBS one at a time for embryonic gut dissection to allow clearer visualization. Keep the other embryos in another petri dish with cold PBS on ice.
4. Dissect the full-length embryonic gut from stomach to hindgut using fine microdissection forceps. Start by removing the head, limbs, and tail, then open the chest and abdominal wall to expose the chest and abdominal cavity. Carefully remove the lung and liver, and locate the stomach, which is just ventral to the liver on the left side of the embryo. Trace the length of the gut tube to locate the anal canal. Extract the embryonic gut by clipping off other tissues away from the gut tube using forceps (*see* **Note 10**).
5. Transfer embryonic gut to individual wells of a 96-well plate containing cold PBS using low-retention wide-bore P1000 pipette tips and keep on ice while dissecting the remaining embryos.

3.2 Embedding the Embryonic Gut Tissue in Matrigel

1. Transfer embryonic gut suspended in 100 μL of PBS to the well of a 12-well plate using low-retention wide-bore P1000 pipette tips. Work on one gut tissue at a time and keep the others in PBS in the 96-well plate on ice (*see* **Note 11**).
2. Use fine forceps to adjust the gut tube orientation and untangle the tissue if needed (*see* **Note 12**).
3. Remove all PBS using a P200 pipette without disrupting the embryonic gut (*see* **Note 13**).
4. Add 20 μL of Matrigel using low-retention P200 pipette tips to the embryonic gut just enough to allow the tissue to attach to the bottom of the well (*see* **Note 14**).

5. Move the 12-well plate with the lid on to a heating pad at low to medium setting to allow Matrigel to polymerize for 2–3 min (*see* **Note 15**).
6. As soon as the Matrigel is polymerized, remove the 12-well plate from the heating pad and add 50 μL of Matrigel on top of the embryonic gut to cover the entire tissue (*see* **Note 16**).
7. Move the 12-well plate with the lid on to the heating pad to allow Matrigel to polymerize for 3–4 min.
8. Continue to the next embryonic gut in another well in the 12-well plate until all gut tissues are covered in Matrigel.

3.3 Culturing of the Embryonic Gut Explant

Perform the following steps in a BSC to prevent contamination of tissue cultures:

1. Keep the 12-well plate containing Matrigel-covered gut tissues in a humidified tissue culture incubator at 37°C, supplied with 5% CO_2 for 10 min to ensure complete polymerization of the Matrigel.
2. Add 1.5 mL pre-warmed explant media to each well (*see* **Note 6**).
3. Culture the explants in a humidified tissue culture incubator at 37°C supplied with 5% CO_2.
4. Change media every 2–3 days.

3.4 Tracing of ENCCs Migration

1. As a baseline, image the embryonic gut explants using an inverted epifluorescence microscope with a cell culture plate holder and incubator chamber after plating of the explant in Matrigel and adding the media (*see* **Note 3**).
2. Images can be taken at different time points. Depending on the required analysis, images can be taken using z-stack to cover the full thickness of the gut tissue if available. To visualize the whole length of the embryonic gut, images can be taken from different objective focuses, and stitching can be applied if available.
3. With the addition of a chemoattractive factor, outward migration of ENCCs can be observed as early as 24 h. Quantification of ENCCs migration can be measured as the length of *Sox10-Venus* positive neurites (Fig. 1).
4. Proceed with the desired downstream analysis.

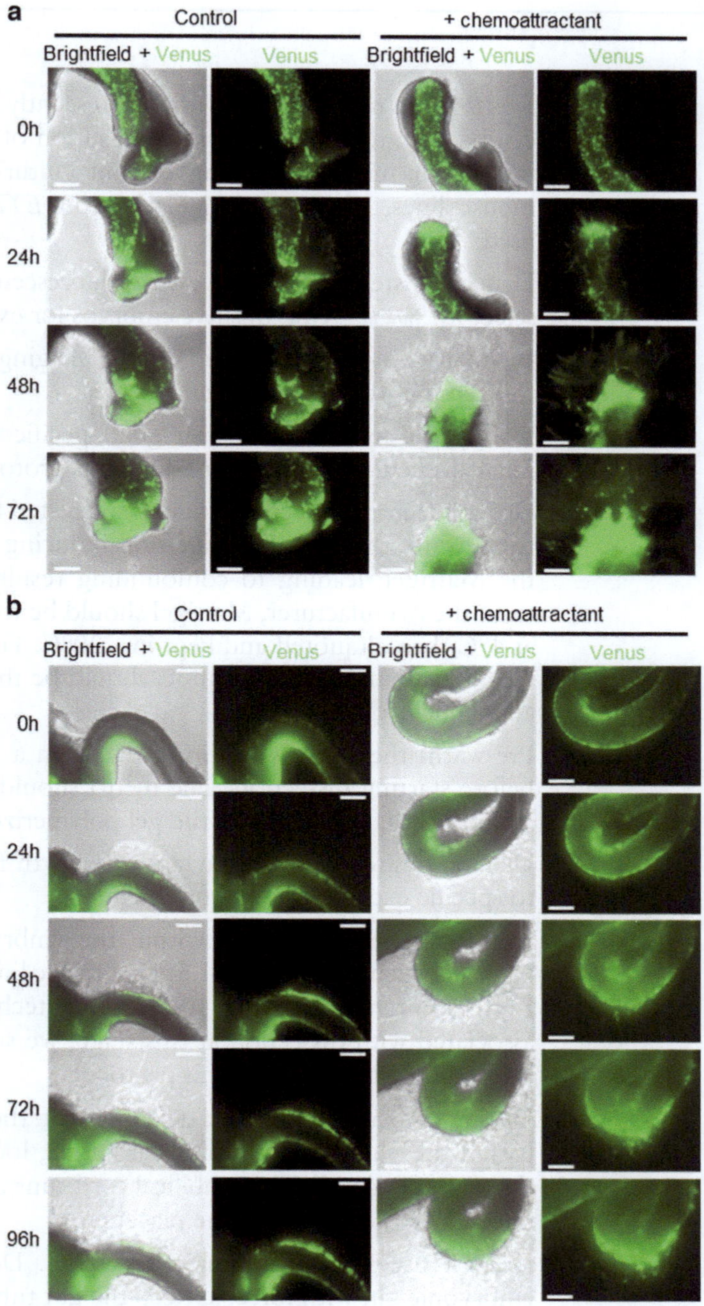

Fig. 1 Representative micrographs of embryonic gut explants. (**a**) E11.5 *Sox10-Venus+* embryonic gut explants cultured in control media or media supplemented with chemoattractive factor. (**b**) E14.5 *Sox10-Venus+* embryonic gut explants cultured in control media or media supplemented with chemoattractive factor. Scale bar: 200 μm (**a**, **b**)

4 Notes

1. *Sox10-Venus* reporter mice with fluorescently labeled ENCCs can be used to allow live-cell visualization of ENCCs and to trace their migration during explant culturing. Alternative mouse lines, such as *Wnt1-Cre; R26R-EYFP,* can also be used.
2. Dissection stereo microscope with fluorescence is needed to select for *Sox10-Venus* positive embryos for explant culturing.
3. Incubator chamber is needed during imaging of the explants to keep the Matrigel polymerized.
4. Embryonic tissues are very fragile; specific fine forceps for microdissection should be used for this protocol.
5. Growth factor-reduced Matrigel is suggested for this protocol to prevent other possible migration-inducing molecules from the Matrigel, leading to confounding results. Upon arrival from the manufacturer, Matrigel should be thawed overnight at 4°C then aliquoted and kept in −20°C. For this protocol, an adequate number of aliquots should be thawed overnight at 4°C before usage.
6. Pre-warm the explant culturing media in a 37°C incubator before starting dissection. The media should be warm upon adding to Matrigel to keep the gel polymerized.
7. Use the desired chemoattractant or growth factor according to specific experimental requirements.
8. Development of the ENS within the embryonic gut takes place between E9.5 and E14.5. Due to the limited number of ENCCs entering the gut at E9.5 and the technical challenges associated with dissection at this stage, we suggest E10.5 as the earliest time point for this protocol.
9. During embryonic intestinal development, the gut tube loops within the physiological umbilical hernia, located outside of the abdominal wall. The umbilical cord runs closely along the intestine and connects to the placenta.
10. Do not use scissors for microdissection. Do not hold the embryonic gut with forceps. Keep the gut tube intact by gently using fine forceps to hold and remove other tissues surrounding it.
11. Transfer and work on one gut at a time to prevent embryonic gut tissue from drying out.
12. Do not use forceps to pick up the gut tissue. Apply light pressure to push tissue around in the PBS droplet.
13. Remove all PBS to prevent diluting Matrigel, as this will affect the solidity of the gel matrix.

14. While working with Matrigel, pipette slowly and prevent the formation of bubbles. Always keep Matrigel on ice until plating to prevent premature polymerization.
15. Do not turn the heating pad to high heat, as it may degrade the gut tissue. Do not leave the 12-well plate with gut tissues on the heating pad too long, as the exposed gut tissue may dry out.
16. If the tissue becomes loose when additional Matrigel is added, remove excess Matrigel to ensure the tissue lies flat to the bottom of the well. Then, move the plate back to the heating pad to allow the Matrigel to solidify before adding more. For bigger tissues (i.e., E13.5 and E14.5), more than 50 μL of Matrigel may be needed to ensure the entire gut tissue is covered.

Conflict of Interest

The authors have no conflicts of interest to declare that are relevant to the content of this chapter.

References

1. Furness JB (2006) The enteric nervous system. Blackwell Publishing, MA
2. Fujiwara N, Nakazawa-Tanaka N, Yamataka A (2019) Animal models of Hirschsprung's disease: state of the art in translating experimental research to the bedside. Eur J Pediatr Surg 29(4):361–367
3. Jessen KR, Saffrey MJ, Burnstock G (1983) The enteric nervous system in tissue culture. I. Cell types and their interactions in explants of the myenteric and submucous plexuses from guinea pig, rabbit and rat. Brain Res 262(1):17–35
4. Metzger M, Bareiss PM, Danker T, Wagner S, Hennenlotter J, Guenther E, Obermayr F, Stenzl A, Koenigsrainer A, Skutella T, Just L (2009) Expansion and differentiation of neural progenitors derived from the human adult enteric nervous system. Gastroenterology 137(6):2063–2073
5. Rauch U, Schäfer KH (2003) The extracellular matrix and its role in cell migration and development of the enteric nervous system. Eur J Pediatr Surg 13(3):158–162
6. Shibata S, Yasuda A, Renault-Mihara F, Suyama S, Katoh H, Inoue T, Inoue YU, Nagoshi N, Sato M, Nakamura M, Akazawa C, Okano H (2010) Sox10-Venus mice: A new tool for real-time labeling of neural crest lineage cells and oligodendrocytes. Mol Brain 3:31
7. Passaniti A, Kleinman HK, Martin GR (2022) Matrigel: history/background, uses, and future applications. J Cell Commun Signal 16(4):621–626

Chapter 8

Embryonic Gut Enteric Neural Crest Cells Derived Neurospheres

Dorothy Lee, Naho Fujiwara and Agostino Pierro

Abstract

Enteric neural crest cells (ENCCs) are progenitor cells in the developing gut that give rise to the enteric nervous system (ENS). Mechanisms that regulate the migration, proliferation, and differentiation of these cells are crucial to the proper morphogenesis of the intestinal neural network during embryogenesis. Despite the relatively low population of these cells within the gut tissue, methods for isolation and *in vitro* culture have been established to enrich this cell population and enable investigations into molecular pathways that govern ENS development. Here, we describe a method to generate ENCC-derived neurospheres from the mouse embryonic gut at embryonic day (E) 13.5 by neural selective media enrichment.

Keywords Enteric nervous system (ENS), Enteric neural crest cells (ENCCs), Neurospheres, Embryonic gut, Neural selective media

1 Introduction

The morphogenesis of the enteric nervous system (ENS) during development depends on the migration, proliferation, and differentiation of enteric neural crest cells (ENCCs) [1]. These processes are regulated by different molecular pathways and mechanisms, which are required to form the complex neural network along the whole intestinal tract. Misregulation of these pathways may lead to gastrointestinal neuropathy, such as Hirschsprung disease [2]. Although the proportion of enteric neural progenitors is low relative to the total cell population in the gut [3], isolating ENCCs from embryonic gut tissues allow us to study these key developmental processes at a molecular level. *In vitro* neural selective culturing conditions have been optimized to enrich this cell population without the need for cell labeling. In contrast to two-dimensional cell culture, which leads to neural differentiation of ENCCs after adherence, three-dimensional

Paul Delgado-Olguín (ed.), *Mouse Embryogenesis*, Methods in Molecular Biology, vol. 3010,
https://doi.org/10.1007/978-1-0716-5126-1_8,

suspension culture of neurospheres helps maintain ENCCs as progenitor cells with higher proliferation ability [4].

An alternative application of this methodology extends beyond the study of ENS morphogenesis. *In vitro* expansion of ENCCs derived from neurospheres has been applied as a potential regenerative medicine for Hirschsprung disease. In this case, transplantation of neurospheres to Hirschsprung animal model showed engraftment and differentiation towards neurons and glia cells [5, 6]. Isolation of enteric neural progenitor cells from the excised colonic tissue from patients with Hirschsprung disease has also been successful and can be maintained and expanded in some culture as neurospheres [7]. In the future, patient-derived neurospheres could be developed to establish a method of autologous transplantation or can be used as a screening platform for alternative treatment modalities.

2 Materials

All reagents and materials should be prepared under sterile conditions to prevent contamination.

2.1 Animals

1. C57BL/6 mice (*see* **Note 1**).

2.2 Equipment

1. Dissection stereo microscope.
2. Fine microdissection forceps (Tips dimension: 0.05 × 0.02 mm) (*see* **Note 2**).
3. Biosafety cabinet (BSC).
4. Tissue culture incubator at 37°C supplied with 5% CO_2.
5. Hemocytometer.

2.3 Tubes and Culture Plates

1. 6 cm petri dishes.
2. Low-retention wide-bore P1000 pipette tips.
3. Low-retention P200 pipette tips.
4. 96-well cell culture plate.
5. 1.5 mL Eppendorf tubes.
6. Ultra-low attachment flask 25 cm^2 (*see* **Note 3**).
7. 15 mL Falcon tubes.

2.4 Reagents

1. Phosphate-buffered saline (PBS).
2. Collagenase 1 (1 mg/mL diluted in Dulbecco's phosphate-buffered saline (DPBS)).
3. Neural selective media (NSM): Dulbecco's modified Eagle's medium/Nutrient Mixture F-12 (DMEM/F-12), 1% penicillin/streptomycin, B27 supplement, N2 supplement, 20 ng/mL

epidermal growth factor (EGF), 20 ng/mL basic fibroblast growth factor (bFGF).

3 Methods

3.1 Embryonic Gut Tissue Dissection

1. Following institutional guidelines and Animal Care Committee approval, euthanize pregnant dam at embryonic day (E) 13.5 and immediately extract the uterus and place in 6 cm petri dish with cold PBS.
2. Extract whole embryos from the uterus under a dissecting stereo microscope. To free the whole embryo, carefully cut the umbilical cord as close to the placenta as possible without disrupting the embryonic gut (*see* **Note 4**). Remove the yolk sac and amniotic sac.
3. Transfer embryos to a new 6 cm petri dish with cold PBS one at a time for embryonic gut dissection to allow clearer visualization. Keep the other embryos in another petri dish with cold PBS on ice.
4. Dissect the full-length embryonic gut from stomach to hindgut using fine microdissection forceps. Start by removing the head, limbs, and tail, then open the chest and abdominal wall to expose the chest and abdominal cavity. Remove the lung and liver carefully and locate the stomach, which is just ventral to the liver on the left side of the embryo. Trace the length of the gut tube to locate the anal canal. Extract the embryonic gut by clipping off other tissues away from the gut tube using forceps (*see* **Note 5**).
5. Transfer the embryonic gut to individual wells of a 96-well plate containing cold PBS using low-retention wide-bore P1000 pipette tips and keep on ice while dissecting the remaining embryos.

3.2 Embryonic Gut Tissue Dissociation and Generation of Neurospheres

Perform the following steps in a BSC to prevent contamination of cell cultures.

1. Transfer each E13.5 embryonic gut suspended in PBS to individual 1.5 mL Eppendorf tubes using low-retention wide-bore P1000 pipette tips.
2. Remove PBS with a P200 pipette.
3. Add 250 μL collagenase 1 to each tube.
4. Incubate in a 37°C tissue culture incubator for 40 min.
5. Manually break down the tissue into a cell suspension by pipetting up and down using low-retention P200 pipette tips.

6. Centrifuge at 2500 RCF for 1 min at room temperature to pellet the cells.
7. Discard the supernatant.
8. Resuspend the cell pellet with 100 μL NSM.
9. Count cells using a hemocytometer.
10. Plate cells in 25 cm^2 ultra-low attachment flasks at a density of 2×10^6 cells/mL in NSM to a final volume of 4 mL each flask (*see* **Note 6**).

3.3 Culturing of Neurospheres

1. Culture cells in a humidified tissue culture incubator at 37°C, supplied with 5% CO_2.
2. Change media every 3–4 days:
   a. Check formation of neurospheres and estimate cell density under the microscope.
   b. Collect all cells and media from flasks into 15 mL Falcon tubes (*see* **Note 7**).
   c. Centrifuge at 500 RCF for 1 min at room temperature.
   d. Carefully remove supernatant with P200 pipette without removing any cells.
   e. Resuspend cells in NSM according to estimated cell density (2×10^6 cells/mL NSM) (*see* **Note 8**).
   f. Plate cells back to the same 25 cm^2 ultra-low attachment flasks.

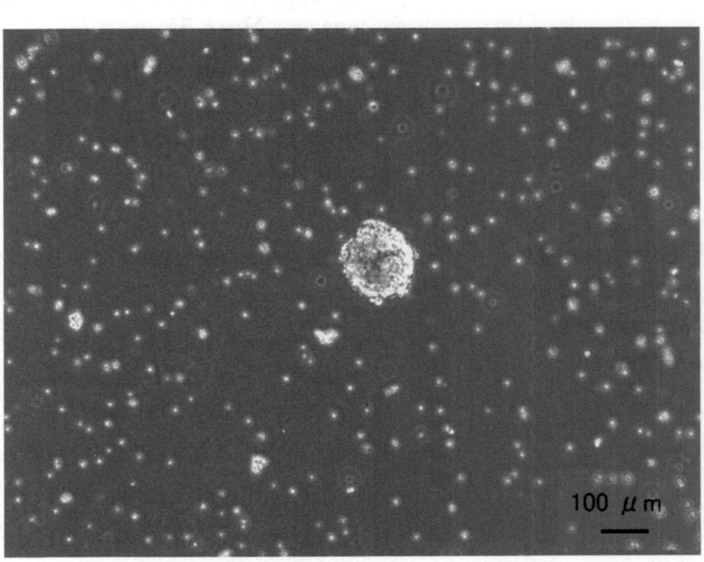

Fig. 1 Representative micrograph of neurospheres derived from E13.5 mouse embryonic gut at 7 days after culturing

3. Neurospheres should form after 7 days in culture (Fig. 1). Proceed with desired downstream analysis.

4 Notes

1. Transgenic reporter mice with fluorescently labeled ENCCs (e.g., *Sox10-Venus* [8]) can be used to allow live cell visualization of ENCCs derived neurospheres to confirm their enteric neural progenitor source or for transplantation tracing purposes.
2. Embryonic tissues are very fragile; microdissection-specific fine forceps should be used for this protocol.
3. Low attachment flasks must be used for neurosphere suspension formation.
4. At E13.5, the mouse embryonic gut loops within the physiological umbilical hernia outside of the abdominal wall, where the umbilical cord runs closely along the intestine and is connected to the placenta.
5. Do not use scissors for microdissection. Do not hold the embryonic gut with forceps. Keep the gut tube intact by holding and removing other tissues surrounding it using fine forceps gently.
6. Ensure cell density is not too low, as an adequate number of cells is needed to encourage neurosphere formation.
7. Some neurospheres will start forming at this stage. Be gentle when collecting cells and media from the flasks, and do not dissociate already formed neurospheres. For easier handling, flasks of the same experimental group can be pooled together.
8. Volumes of NSM used to resuspend cells shall be adjusted according to the cell density observed before media change. If cell density is too low, cells can be pooled together and re-plate into fewer flasks. On the other hand, if the cell density is too high, split the cells into more flasks.

Conflict of Interest

The authors have no conflicts of interest to declare that are relevant to the content of this chapter.

References

1. Nagy N, Goldstein AM (2017) Enteric nervous system development: a crest cell's journey from neural tube to colon. Semin Cell Dev Biol 66:94–106
2. Heanue TA, Pachnis V (2007) Enteric nervous system development and Hirschsprung's disease: advances in genetic and stem cell studies. Nat Rev Neurosci 8(6):466–479
3. Drokhlyansky E, Smillie CS, Van Wittenberghe N, Ericsson M, Griffin GK, Eraslan G, Dionne D, Cuoco MS, Goder-Reiser MN, Sharova T, Kuksenko O, Aguirre AJ, Boland GM, Graham D, Rozenblatt-Rosen O, Xavier RJ, Regev A (2020) The human and mouse enteric nervous system at single-cell resolution. Cell 182(6):1606–1622
4. Reynolds BA, Weiss S (1992) Generation of neurons and astrocytes from isolated cells of the adult mammalian central nervous system. Science 255(5052):1707–1710
5. Cooper JE, McCann CJ, Natarajan D, Choudhury S, Boesmans W, Delalande JM, Vanden Berghe P, Burns AJ, Thapar N (2016) In vivo transplantation of enteric neural crest cells into mouse gut; engraftment, functional integration and long-term safety. PLoS One 11(1):e0147989
6. Hotta R, Cheng LS, Graham HK, Pan W, Nagy N, Belkind-Gerson J, Goldstein AM (2016) Isogenic enteric neural progenitor cells can replace missing neurons and glia in mice with Hirschsprung disease. Neurogastroenterol Motil 28(4):498–512
7. Wilkinson DJ, Bethell GS, Shukla R, Kenny SE, Edgar DH (2015) Isolation of enteric nervous system progenitor cells from the aganglionic gut of patients with Hirschsprung's disease. PLoS One 10(5):e0125724
8. Hibata S, Yasuda A, Renault-Mihara F, Suyama S, Katoh H, Inoue T, Inoue YU, Nagoshi N, Sato M, Nakamura M, Akazawa C, Okano H (2010) Sox10-Venus mice: a new tool for real-time labeling of neural crest lineage cells and oligodendrocytes. Mol Brain 3:31

Chapter 9

Preparation of Mouse Embryonic Eye Sections for Routine Histology

Esmeralda Lira-Romero, Francisco Javier Estrada-Mena and Israel Ramírez-Sánchez

Abstract

Eye development is a complex process that begins during the first stages of embryonic development; during its formation, a series of cell proliferation, migration, and differentiation events occur. A frequently used laboratory method is tissue staining with two dyes known as hematoxylin–eosin (H&E). H&E staining allows us to observe the arrangement of the cell layers and the formation of characteristic structures. It is fundamental for research in developmental biology since it facilitates a comprehensive analysis of the formation of organs and systems.

In this chapter, we will describe the procedure for acquiring H&E-stained images of the embryonic eye structure, specifically focusing on the retina. This technique serves as a fundamental tool for understanding ocular morphogenesis, organogenesis, and structural modifications resulting from experimental interventions.

Keywords Mouse embryonic eye, Retina, Histological sections, Hematoxylin–eosin stain, Paraffin, Perfusion, Descaling

1 Introduction

The retina, a crucial component of the visual system, develops from the anterior neural plate during embryogenesis. The retina is located in the innermost part of the eyeball, and its function is to transform light into a nervous impulse that reaches the brain through the optic nerve. Thus, converting light into the images we perceive [1].

Some ocular disorders, such as anophthalmia, microphthalmia, and retinitis pigmentosa, have their origin in abnormalities in the development of the retina. Understanding the mechanisms underlying these alterations from embryonic stages can provide important information for their diagnosis, prevention, and possible treatment [2, 3].

Paul Delgado-Olguín (ed.), *Mouse Embryogenesis*, Methods in Molecular Biology, vol. 3010,
https://doi.org/10.1007/978-1-0716-5126-1_9,

A frequently employed procedure for analyzing cellular structures requires tissue staining with H&E. The process involves: (1) fixation of the tissue, either by perfusion or immersion, to avoid degradation of the cellular components; (2) inclusion in paraffin that provides support and facilitates making fine sections of the tissue; and (3) staining with hematoxylin, which stains the tissue with dark blue nucleic acids, and eosin that provides the pink to red blue of the cytoplasm and extracellular matrix [4–7].

H&E staining has been used for at least a century and remains an essential technique in histology to study various tissue types and their morphological changes [8, 9]. In the context of ocular development, this methodology provides important information about its normal formation and the possible detection of structural alterations during its development, as well as changes produced by experimental interventions.

2 Materials

Prepare all solutions using ultrapure water (prepared by purifying deionized water) and analytical-grade reagents. Prepare and store all reagents at room temperature, unless indicated otherwise. Procedures were developed following laboratory animal care and biological waste disposal regulations.

2.1 Anesthesia

1. Ketamine (100 mg/1 mL).
2. Xylazine (2%).
3. 1 mL insulin syringes.
4. Ethanol: 70% solution in water.

2.2 Fixation of Embryos

1. Surgical kit (autoclaved).
2. Curved scissors.
3. Vinyl gloves.
4. Protective clothing.
5. Eyeglasses and a mask.
6. Fume extraction hood.
7. Plastic bags, paper towels.
8. 10 mL syringes.
9. Absorbent pad.
10. Phosphate-buffered saline (PBS; 1×): 137 mM NaCl, 2.7 mM KCl, 10 mM Na_2HPO_4, and 1.8 mM KH_2PO_4, pH 7.4 (*see* **Note 1**).
11. Paraformaldehyde 4% solution (w/v) in PBS 1×: Add 4 g paraformaldehyde to 50 mL of H_2O. Add 1 mL of 1 M NaOH and stir gently on a heating block at ~60°C until the

paraformaldehyde is dissolved. Add 10 mL of 1× PBS and allow the mixture to cool to room temperature. Adjust the pH to 7.4 with 1 M HCl, then adjust the final volume to 100 mL with H_2O. Filter the solution through a 0.45 μm membrane filter to remove any particles (*see* **Note 2**).

2.3 Obtaining Embryos

1. Surgical kit (autoclaved).
2. Clean Petri dishes.
3. Scalpel blades and handle.
4. Dark glass sealable bottles.

2.4 Descaling

1. Ethylenediaminetetraacetic acid (EDTA) 10% solution (w/v): Add 10 g of EDTA to 80 mL of double-distilled water.
2. Dark glass sealable bottles.

2.5 Paraffin Embedding

1. Phosphate-buffered saline (PBS; 1×): 137 mM NaCl, 2.7 mM KCl, 10 mM Na_2HPO_4, and 1.8 mM KH_2PO_4, pH 7.4.
2. Xylene.
3. Disposable plastic cassettes for histology (with lids).
4. Paraffin for histology, melting point 56–57°C previously melted.
5. Automated tissue processor.
6. Ethanol for dehydration series: 70, 80, 90, 96, 100, 100% solution in water.
7. Ethanol 100%-Xylene.
8. Ethanol for hydration series: 100, 96, 70%.
9. Metal molds for inclusion.
10. Small forceps.

2.6 Sectioning

1. Rotary microtome.
2. Tissue water bath with a thermometer.
3. Disposable high-profile microtome blades.
4. Fine paintbrushes to remove paraffin from paraffin sections.
5. Fine paintbrushes to pick up sections.
6. Coated glass slides.

2.7 Hematoxylin–Eosin Staining

1. Laboratory stove that reaches 60°C.
2. Coping with glasses.
3. Gill's Hematoxylin Solution: Place 750 mL of double-distilled water in a beaker. Add 250 mL of ethylene glycol (C2H6O2) and mix with constant stirring for 30 min. Add 2 g of anhydrous hematoxylin powder and stir for an additional 30 min period. Add 0.2 g of sodium iodate ($NaIO_3$) and continue stirring for

30 min. Add 17 g of aluminum sulfate ($Al2(SO4)3$) and let the solution mix for 3 h. Finally, add 0.2 mL of glacial acetic acid ($CH3COOH$) and stir for an additional hour.

4. Filter the solution through a 0.45 μm membrane filter (*see* **Note 3**).
5. Eosin: place 10 g of eosin in 200 mL of double-distilled water, then add 800 mL of ethanol. Mix until dissolved. Filter the solution through a 0.45 μm membrane filter (*see* **Note 4**).

3 Methods

3.1 Fixation of Embryos in Utero

1. Anaesthetize the pregnant mouse intraperitoneally, using appropriate pharmacological agents (*see* **Note 5**).
2. Establish a clean work area, preferably in a fume extraction hood.
3. Once the pregnant mouse is deeply anaesthetized, carefully place it on an absorbent pad in a supine position and secure the upper extremities (*see* **Note 6**).
4. Clean the abdominal area with 70% ethanol and make a "V"-shaped incision in the abdomen using curved scissors, to expose both the abdominal and thoracic cavities.
5. Cut through the diaphragm and sternum to expose the heart (Fig. 1).

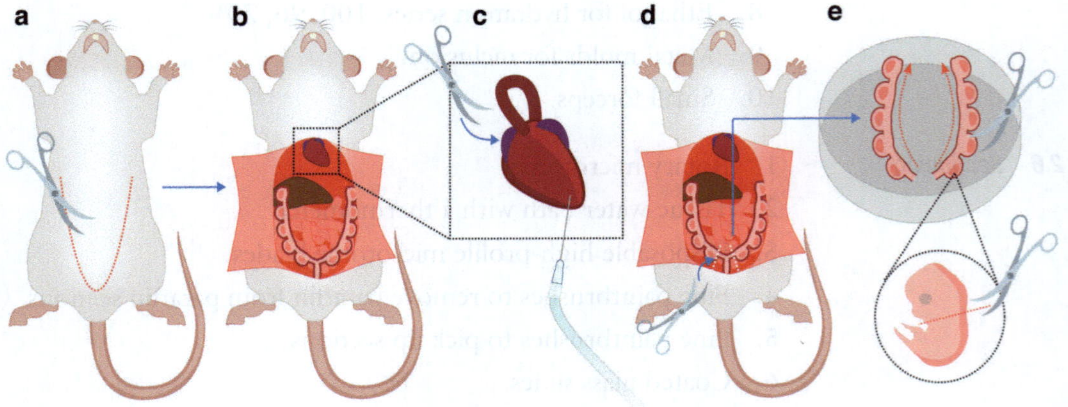

Fig. 1 Diagram illustrating the surgical procedure for in utero embryo fixation in a pregnant mouse. (**a**) Initial midline abdominal incision . (**b**) Exposure of the abdominal and thoracic cavities, with visualization of the internal organs. (**c**) Close-up of the cardiac cavity; scissors point to the incision site in the right atrium, while the needle indicates the entry point into the left ventricle to initiate transcardiac perfusion. with PFA 4%. (**d**). Excision of the uterine horns. (**e**). Identification of the placental sacs; both the cutting plane for embryo collection and the cephallic region intended for eyeball post-fixation are indicated. Arrows indicate the chronological sequence of procedure.

6. Before introducing the fixative into the blood vessel system, it is necessary to remove the blood first, to do this, first circulate the 1× PBS.
7. Introduce the syringe needle into the left ventricle and begin the perfusion. Then, proceed to cut the right atrium to allow both blood and fixative to escape from the circulation (*see* **Note** 7).
8. Once the liver and placental sacs acquire a uniform pale color, remove the syringe containing 1× PBS and insert the syringe with 4% paraformaldehyde (4% PFA) in its place (*see* **Note 8**).

3.2 Obtaining Embryos

Once the perfusion has been performed:

1. Cut the bicornuate uterus that contains the embryos and place it in a Petri dish.
2. Cut the entire length of the placental sacs to release the embryos.
3. Once released, section the head with a scalpel (Fig. 1).
4. Place the embryos in clean glass vials with sufficient 4% PFA and post-fixate them overnight at 4°C (Fig. 2).

3.3 Descaling

1. Immerse the tissue in a 10% EDTA solution for 5 days to reduce bone hardness (*see* **Note 9**).

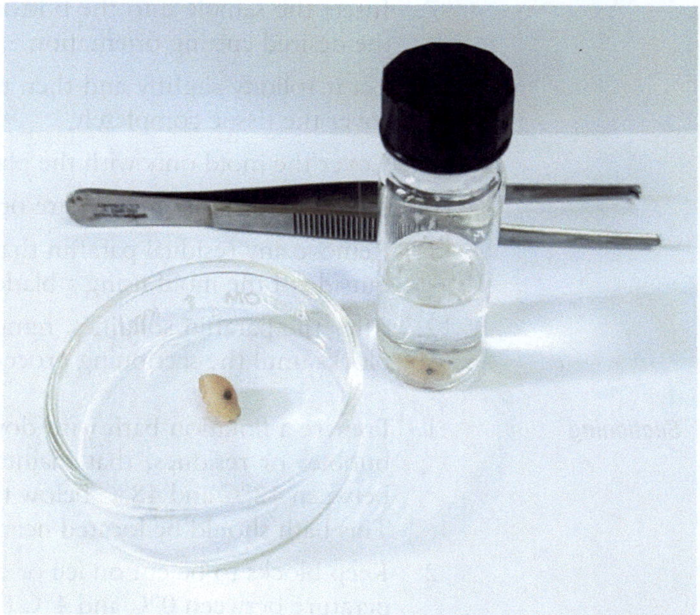

Fig. 2 Post-fixation process of the embryonic mouse eye at E 18.5. The image illustrates the heads of embryos in a petri dish and immersed in 4% paraformaldehyde.

2. Place the vials with the tissue on a shaker at room temperature and shake to ensure uniform distribution of the solution.
3. Change the EDTA solution every 24 h.
4. After decalcification, wash the samples with tap water for at least 1 h to remove EDTA excess and proceed with paraffin embedding.

3.4 Paraffin Embedding

1. Wash the tissues four times with 1× PBS for 15 min each with agitation.
2. Subsequently, place the tissues in embedding cassettes, assigning one head per cassette.
   The following steps can be performed manually or in an automated tissue processor.
3. Dehydrate the tissues by immersing the cassettes in the following series of alcohols: 70, 80, 90, 96, 100. 100% for 1 h each, at room temperature (*see* **Note 10**).
4. Immerse the cassettes with the tissue in three series of Xylene for 1 h at room temperature (*see* **Note 11**).
5. Infiltrate the tissue with three series of melted paraffin, immersing the cassettes with the tissues at 50°C for 1 h for each series. After infiltration, remove the tissues from the cassette with forceps.
6. Place a small amount of melted paraffin at the bottom of a metal inclusion mold.
7. Insert the sample into the bottom of the mold according to the desired cutting orientation.
8. Let it solidify slightly and then pour liquid paraffin again to cover the tissue completely.
9. Cover the mold only with the plastic cassette (*see* **Note 12**).
10. Let cool to room temperature or on a cold plate.
11. Remove any residual paraffin that may have remained on the outside of the mold using a blade.
12. After the paraffin solidifies, remove the mold, and store the blocks until the sectioning procedure (*see* **Note 13**).

3.5 Sectioning

1. Prepare a flotation bath with double-distilled water, without bubbles or residues, that maintains a constant temperature between 42°C and 48°C below the melting point of paraffin. This bath should be located near the microtome.
2. Keep blocks to be cut on ice or cool them at a constant temperature between 0°C and 4°C (*see* **Note 14**).
3. Place the block containing the sample in the block holder of the microtome arm and adjust its orientation using the controls on the block holder (*see* **Note 15**).

4. Place the blade in the microtome blade holder and secure it with the clamping lever. Place the safety bar to avoid cutting yourself (*see* **Note 16**).
5. Adjust the blade tilt angle, typically between 5° and 10°, which determines the angle at which the blade edge touches the sample. Modify according to necessity.
6. Remove the safety bar from the blade and bring it as close to the sample as possible without touching it. Begin roughing to have the sample on the surface, with thicknesses between 10 and 30 μm normally. Once you have reached the tissue area of interest, adjust the thickness of the cut (*see* **Note 17**).
7. Polish the face of the block by cutting a few thin sections.
8. Obtain the cuts into straight strips (*see* **Notes 18 and 19**).
9. Carefully lift the sections with the smooth (shiny) side down and place them on the water surface in a gentle sweeping motion.
10. Leave the sectioned samples on the water surface long enough to flatten.
11. Immerse a gelatin-pretreated or silanized microscope slide to take the sections.
12. Slowly remove the slide from the flotation bath in an upright position to prevent the sections from falling.
13. Let the section dry at room temperature for 10 to 30 min.

3.6 Hematoxylin–Eosin Staining

Dewaxing of the tissue sections

1. Place the slides you wish to dye in the oven at 57°C, inside a metal basket, for 30 min.
2. Remove the slides from the oven and clean the paraffin from the samples with two changes of xylene, the first for 10 min and the second for 5 min (*see* **Note 20**).

Hydration

1. Hydrate the tissue with the following alcohol concentration series:
   Ethanol-Xylol 5 min.
   100% ethanol for 5 min.
   Ethanol 96% for 5 min.
   Ethanol 70% for 5 min.
2. Rinse the slides with distilled water at room temperature, making three changes of 5 min each.

Staining

1. Immerse the slides in a glass Coplin bottle containing Gill's hematoxylin and shake it for 5 min.
2. Remove the hematoxylin and wash the slides under running water (tap water) at room temperature for at least 3 min to remove excess hematoxylin.

3. Perform two changes with double-distilled water for 3 min each.
4. Stain the samples in eosin Y solution for 5 min.

Dehydration

1. Dehydrate the samples with the following alcohol series:
   Ethanol 70% for 2 min.
   Ethanol 80%-Xylol 2 min.
   Ethanol 96% for 2 min.
   100% ethanol for 2 min.
   100% Ethanol-Xylol 2 min.
2. Clean the samples with two changes of xylene for 2 min per change.

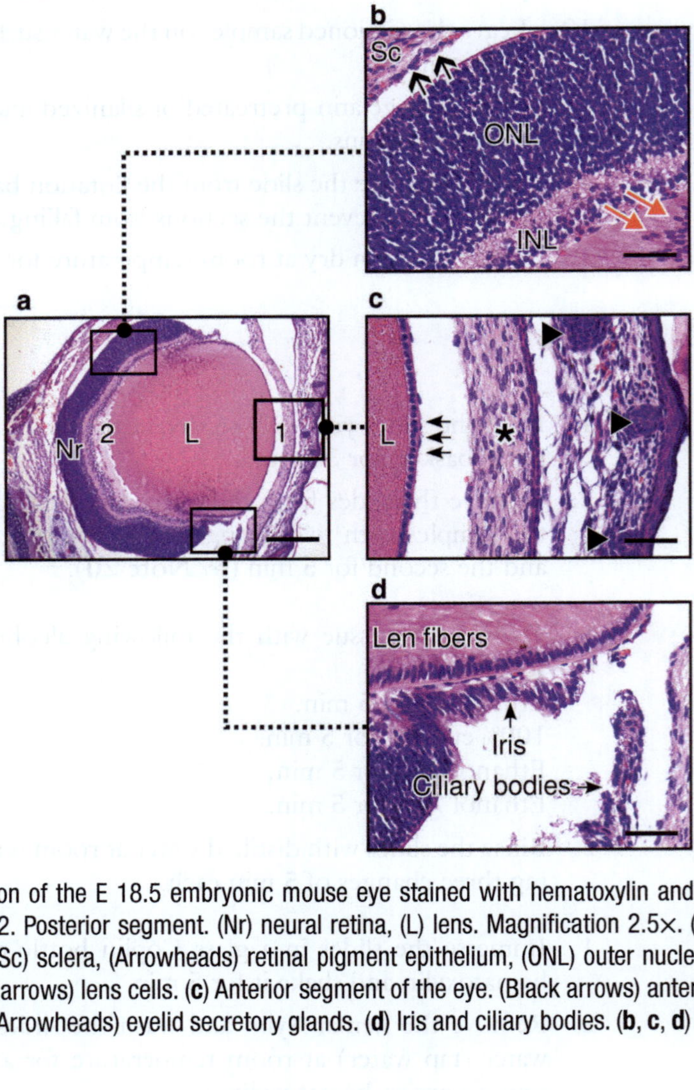

Fig. 3 Cross-section of the E 18.5 embryonic mouse eye stained with hematoxylin and eosin (H&E). (**a**) 1. Anterior segment. 2. Posterior segment. (Nr) neural retina, (L) lens. Magnification 2.5×. (**b**) Magnification of the neural retina. (Sc) sclera, (Arrowheads) retinal pigment epithelium, (ONL) outer nuclear layer, (INL) inner nuclear layer, (Red arrows) lens cells. (**c**) Anterior segment of the eye. (Black arrows) anterior lens epithelium, (Asterisk) cornea, (Arrowheads) eyelid secretory glands. (**d**) Iris and ciliary bodies. (**b**, **c**, **d**) Magnification 40×. Scale bar 50 μm

3.7 Mounting

1. Remove the slides one by one, add one or two mounting medium drops on the tissue and place a coverslip.
2. Let the slides dry at room temperature.
3. Observe under the microscope and obtain images under the bright-field microscope (Fig. 3).

4 Notes

1. Sterilize it and keep it at 4°C. Use it cold when infusing.
2. Use it cold when infusing.
3. The staining solution can be used immediately or stored at room temperature in a dark glass resealable bottle. This solution has a longer shelf life than other types of hematoxylins.
4. Store at room temperature in a dark glass-sealable bottle.
5. The dose may vary depending on gestational age, body weight, and individual sensitivity. We suggest using: Ketamine (100 mg/1 mL)-Xylazine 2%. In a 70:30 ratio (1 μL/1 g of body weight).
6. To ensure that the mouse has reached a deep sedation plane, apply pressure to the base of the tail or observe the absence of the plantar reflex [10].
7. It is important to maintain a constant flow.
8. Circulate approximately 10 mL of 1× PBS and fixative.
9. Bone hardness reduces as decalcification progresses. Perform periodic hardness tests on the samples, using a needle, or similar instrument to check the ease of penetration into the tissue [11].
10. Dehydration must be complete; otherwise, it will affect the subsequent infiltration of the paraffin and subsequent processing steps will fail.
11. This step is essential to eliminate any alcohol residue from the previous process and to eliminate fat, which could cause the paraffin block to soften, making it difficult to cut later.
12. It is suggested to include the sample identification number.
13. Paraffin tissue blocks can be stored at room temperature for long periods [4].
14. This makes the paraffin harder and, therefore, easier to cut.
15. If the sample to be cut is very small, before starting to cut, it is suggested to delimit the cutting area. This is achieved by making a pyramid in the sample with the help of a knife.

This procedure allows obtaining straight strip sections with the appropriate dimensions [12].

16. Previously, lock the right-side crank by placing the microtome lock to prevent the arm from moving down and potentially damaging the paraffin block or injuring the operator.
17. The lower limit for most paraffin-embedded tissues is 3 μm thick; however, a thickness of 5 μm is recommended for eye sections.
18. The blade should make smooth contact with the block to produce thin and uniform cuts.
19. Use a fine wet brush or tweezers to pick up sections of either the blade or the block face.
20. Place the slides in a Coplin glass, or as many as necessary.

Conflict of Interest

The authors have no conflicts of interest to declare that are relevant to the content of this chapter.

References

1. Graw J (2010) Eye development. Curr Top Dev Biol 90:343–386
2. Stenkamp DL (2015) Development of the vertebrate eye and retina. Prog Mol Biol Transl Sci 134:397–414
3. Glaser T, Jepeal L, Edwards JG, Young SR, Favor J, Maas RL (1994) PAX6 gene dosage effect in a family with congenital cataracts, aniridia, anophthalmia and central nervous system defects. Nat Genet 7(4):463–471
4. Suvarna KS, Layton C, Bancroft JD (2018). Bancroft's theory and practice of histological techniques (e-book). Elsevier Health Sciences, Amsterdam
5. Eltoum I, Fredenburgh J, Myers RB, Grizzle WE (2001) Introduction to the theory and practice of fixation of tissues. J Histotechnol 24(3):173–190
6. Sadeghipour A, Babaheidarian P (2019) Making formalin-fixed, paraffin embedded blocks. Methods Mol Biol 1897:253–268
7. Luna LG (1968). Manual of histologic staining methods of the Armed Forces Institute of Pathology, 3rd edn. McGraw-Hill, New York
8. Al-Taee RAM, AL-Aameli MH, Al-Qazwini YM (2019) Histological techniques: a brief historical overview. J Glob Sci Res 2:218–223
9. Coleman R (2006) The long-term contribution of dyes and stains to histology and histopathology. Acta Histochem 108(2):81–83
10. Close B, Banister K, Baumans V, Bernoth EM, Bromage N, Bunyan J, Erhardt W, Flecknell P, Gregory N, Hackbarth H, Morton D, Warwick C (1997) Recommendations for euthanasia of experimental animals: Part 2. DGXT of the European Commission. Lab Anim 31(1):1–32
11. Mirando AJ, Hilton MJ (2021) Demineralized murine skeletal histology. Methods Mol Biol 2230:283–302
12. Sadeghipour A, Babaheidarian P (2019) Making formalin-fixed, paraffin embedded blocks. Methods Mol Biol 1897:253–268

Chapter 10

Embryonic Mouse Cardiac Fibroblast Isolation

Enrique Mendez-Bolaina, Elena de la Cruz Herrera-Cogco and Israel Ramírez-Sánchez

Abstract

Mouse cardiac fibroblasts (FBs) have been widely used as in vitro models for studying fundamental biological processes and mechanisms underlying cardiac pathologies, as well as identifying potential therapeutic targets. Cardiac FBs preparations are relatively easy to culture in a dish and can be manipulated using molecular and pharmacological tools. Because FBs rapidly decrease cell cycle division and proliferative rate after birth, they are prone to undergo phenotypic changes and senescence in cell culture as soon as a few passages. Therefore, primary cultures of differentiated FBs with amplified cell cycle from embryonic animals are required for accurate and reliable experiments. Below, we will describe a method that provides good cell yield, quality, and viability of mouse embryonic cardiac fibroblasts (MEFs) primary cultures from E16 CD-1 mice.

Keywords Cardiac fibroblasts, Heart, Enzymatic digestion, Embryonic mice, Isolation, Cell culture

1 Introduction

Seminal studies on specific cardiac regions demonstrated that mouse heart consists of approximately 30% cardiac myocytes and 70% non-myocytes being fibroblast most of these cells [1].

The definition of the fibroblast is based on morphological characteristics (elongated fiber shape) that can vary with location within the organism, as well as the overall activity level of the tissue itself. Morphologically, fibroblasts are flat, spindle-shaped cells that lack a basement membrane [2] and feature a round, elongated nucleus surrounded by vast endoplasmic reticulum [3]. There are different theories regarding the diverse origins of the cardiac fibroblast lineage. In the mammalian embryo, cardiac fibroblasts are thought to be derived primarily from the proepicardial organ via epithelial-to-mesenchymal transition (EMT) [4, 5], although it has been proposed there may be some subsequent contribution from bone marrow-derived precursors [6–8]. In the adult mammalian, it is not well-established how the cardiac fibroblast population is maintained

Paul Delgado-Olguín (ed.), *Mouse Embryogenesis*, Methods in Molecular Biology, vol. 3010,
https://doi.org/10.1007/978-1-0716-5126-1_10,

during normal homeostasis, but it is generally thought that resident fibroblasts and epithelial cells undergoing EMT are the primary source of the low level of turnover within the adult heart [5]. To date, fibroblasts have been difficult to accurately identify. In some cases, fibroblasts are identified based on their spindle shape combined with positive staining for the mesenchymal marker vimentin and the absence of staining for epithelial or other cell types, such as smooth muscle cells, astrocytes, or hematopoietic cells [9].

Cardiac fibroblasts have been termed "sentinel" cells, as they can sense changes in chemical, mechanical, and electrical signals in the heart, and trigger an appropriate response. One of the primary functions of the cardiac fibroblast is the regulation of synthesis and degradation of the extracellular matrix (ECM) in the myocardium to provide a three-dimensional network for myocytes and other cells of the heart to ensure proper cardiac form and function; furthermore, cardiac FBs are involved in synthesis and secretion of growth factors and cytokines into the organ [9]. In fact, under physiological condition cardiac FBs not only produce and remodel the ECM in response to different physiological cues [4], but also are involved in several tissue functions such as regulation of growth during development and proliferation of the cardiomyocyte [10]. Also, FBs directly connect to cardiomyocytes via connexins [11] *and* are able to electrically isolate various portions of the conduction system in the heart [9] and secrete factors that signal to cardiomyocytes in a paracrine fashion [10]. Meanwhile, under pathological conditions, FBs can induce cardiomyocyte hypertrophic and fibrotic responses to injury in the adult heart by secreting cytokines (TNFα, pro-inflammatory interleukins, and TGFβ), active peptides (angiotensin II, endothelin 1) and growth factors [2, 9, 12], which stimulate the myocardium in autocrine and/or paracrine fashions [13].

Hence, cardiac FBs play an important role in normal and pathologic performance of the heart. An understanding of the FBs biology requires in-depth studies using a stable and reliable system in which the biological responses of cardiac fibroblasts to different stimuli can be determined [14]. Thus, the deeper our understanding of fibroblasts within a given environment, the more capable we will be to modulate their responses [7]. Given their several roles within the heart, cardiac FBs provide an excellent model system to study many aspects of heart function and pathophysiology, including the regulation of cell growth, survival, inflammation, hormone secretion, and metabolism [15]. To unravel the molecular and cellular mechanisms for normal cardiac function and cardiac pathology, high-quality isolated cardiac fibroblasts are essential [16, 17]. Although there are several methods to obtain the fibroblast of different animal species there is no consensus on the optimal protocol to isolate and culture a great number of viable cells [17]. However, at their core, these protocols generally all rely on digestion of the heart with an enzymatic solution that will disrupt the hardest tissue releasing into the media the cellular component of the heart [16, 17].

Nevertheless, the culture of these isolated FB's is an important yet difficult point of the process, because as primary cells, they cannot be cultured indefinitely due to the onset of replicative senescence or aneuploidization, leaving just a few passages and little time to set up experiments and get results. Therefore, new techniques or improved protocols need to be established to extend the life span of the cultures [16–18].

Isolation of embryonic FBs now provides better yields and longer cellular viability. Jonsson et al., in their studies, defined the transcriptomic, morphologic, and epigenetic differences between fetal and adult human cardiac fibroblast (fHCFs and aHCFs) by performing a thorough characterization of primary fetal and adult human ventricular CFs using cellular, molecular, and genome-wide sequencing approaches [11, 19]. They showed that fHCFs are smaller and proliferate more quickly than their adult counterpart. Moreover, gene expression profiling revealed higher expression of regulators of heart developent and growth, whereas aHCFs express genes involved in maintaining a mature state [11, 19, 20]. Such differences between adult and fetal cells might influence the viability of the cultures and make embryonic cardiac fibroblasts more desirable to isolate and perform molecular and pathophysiology studies of the heart.

Also, the earliest embryonic FBs have been used as feeder cells to maintain cultures of stem cell, given the fact that embryonic FBs secrete a large amount ECM which directly contributes to the quality of the culture and keeps the stem cells in an undifferentiated state [21, 22]. Another reason to isolate embryonic FBs is that they can be transdifferentiated into a variety of more complex mature cells, such as cardiomyocytes [22, 23], providing different models to study development or cardiac diseases.

This protocol is useful for embryonic mouse cardiac fibroblast isolation, and its primary aim is to maintain a sufficient number of viable cells to study cell signaling, canonical, and non-canonical pathways that could potentially lead to cardiac fibrosis: one of the main features of several cardiac pathologies, which is promoted by FBs [4, 10].

2 Materials

Prepare all solutions using ultrapure water (prepared by purifying deionized water) and analytical grade reagents. Prepare and store all reagents at room temperature, unless indicated otherwise. Follow laboratory animal care and biological waste disposal regulations at all times.

1. ADS buffer preparation: 1 L enzyme-free ADS buffer stock solution. Add about 300 mL ultrapure water to a 1 L glass beaker. Weigh enough amount to prepare 1 L of 20 mM HEPES (free acid) and transfer to a beaker to start dissolving, 0.8 mM $MgSO_4$ (anhydrous), 1 mM NaH_2PO_4, 116 mM NaCl, 180.16 mM D--Glucose, 5.4 mM KCl, 0.02 g phenol red indicator and 1% of penicillin-streptomicin-fungizone (PSF) antibiotic-antimycotic mixture solution (*see* **Note 1**). Adjust pH to 7.35 with NaOH. Set the volume to 1000 mL with ddH_2O and mix to dissolve all salts and reagents. Filter buffer solution inside the hood to avoid contamination and store in a labeled sterile bottle at 4°C (*see* **Note 2**). Enzyme mixture must be prepared on the day and filtered: (a) Ads buffer for 100 mL; (b) Collagenase 0.625% (w/v) 0.0625 g (~130 U/mL); (c) Pancreatin 0.6% (w/v) 0.060 g. Place the mixture on the stirring plate for 3 h to dissolve the enzymes. Once the enzyme mixture is completely dissolved, sterilize by filtration, and then the buffer is ready to use. Enzyme activity may vary depending on the source. If you experience trouble with digestion, the enzyme mixture should be titrated to the optimal.
2. Surgical kit (autoclaved).
3. 50–100 mL spinner flasks (autoclaved).
4. 37°C water bath.
5. Hot stirring plate.
6. 2 × 10 cm plates with enzyme-free ADS buffer.
7. 1 × 10 cm plates.
8. Fetal bovine serum (FBS).
9. Dulbecco's Modified Eagle's Media (DMEM) complete media (10% FBS + 1% PSF).

3 Methods

3.1 Fibroblast Isolation

1. CD-1 mice need to be time-mated using a vaginal plug to designate the embryonic stage. On day 20, sedate the pregnant mice with isoflurane (0.5%) using a gas chamber until recumbent. Shave the abdominal region and clean it with 70% ethanol, open it with scissors to collect the whole litter (8–10 mouse embryos). (Fig. 1a, b).
2. Under a stereoscopic microscope proceed to open the embryo chest using scissors, localize the hearts, and excise tissue, then remove aortas and atrial tissue, and cut the heart into small pieces. Place cardiac tissue into a 10 cm cell culture dish with enzyme-free ADS media to eliminate blood excess and clots. Transfer to a 6 cm plate with enzyme-free ADS buffer.

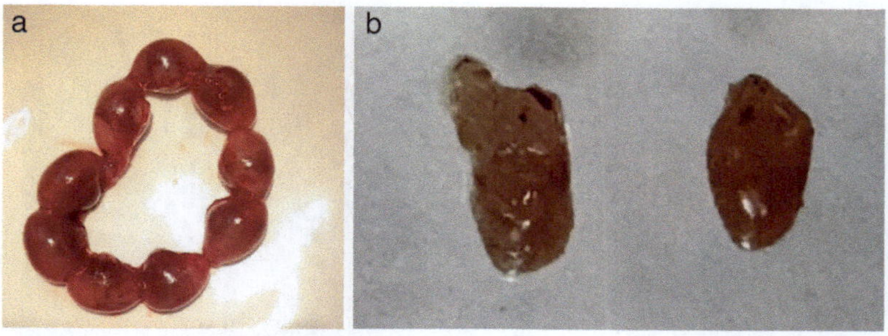

Fig. 1 **(a)** A circular arrangement of red, glossy, grape-like clusters of mouse embryo. CD-1 mice were sedated for embryo collection. **(b)** Mouse embryonic hearts. Embryonic mice were opened to remove the hearts

3. Transfer cardiac tissue portions into a spinning 50–100 mL flask with 10 mL of fresh active ADS buffer (Type II collagenase and pancreatin) over a stir plate with water bath at 37°C for 5 min to digest tissue. Using a 2 mL pipette, carefully discard the supernatant.
4. Add 5 mL of fresh active ADS buffer to the spinning flask and incubate in a water bath for 10 min at 37°C (Fig. 2). After digestion, transfer all the recovered supernatant into a 15 mL Falcon tube with 1 mL of FBS to stabilize cells (*see* **Note 3**). Centrifuge for 5 min at 1500 × *g*, discard the supernatant and resuspend the pellet in 1 mL FBS; place it into the incubator at 37°C (*see* **Note 4**).
5. Repeat the last step five times or until the tissue becomes watery. Each 10 min digestion, recover the pellet, and resuspend it on 1 mL FBS. After five incubations, combine all resulting pellets into a single pool in a 50 mL Falcon tube.

3.2 Plating Cells

1. Centrifuge the pooled cells for 5 min at 1500 × *g*, discard supernatant.
2. Resuspend the pool pellet by adding 7 mL of DMEM complete media. Centrifuge at 1500 × *g* for 5 min.
3. Resuspend pellet with complete DMEM media and plate onto 1 × 10 cm dish. Incubate for 1 h at 37°C in 5% CO_2.
4. After 1 h of incubation, get dish out of incubator, turn the dish clockwise, replace the media, and then incubate 37°C in 5% CO_2.

The next day, shake the plate gently to strip myocytes off; at this time, FBs must be attached to the dish, forming a layer. Replace the media every day, wait until cells reach confluency, and maintain cell culture under standard conditions (Fig. 3).

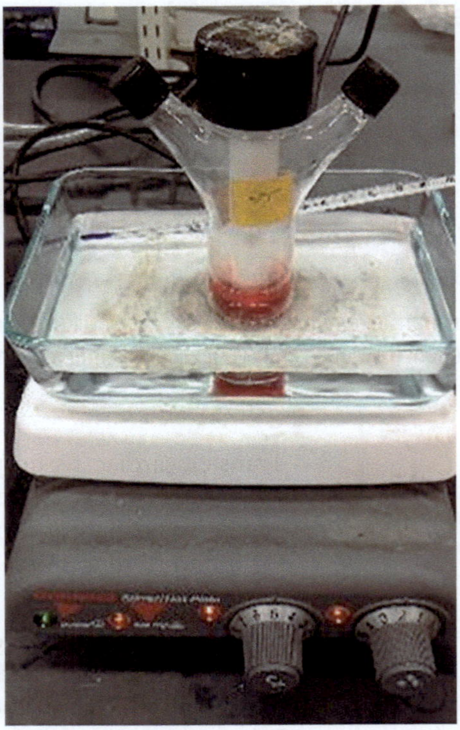

Fig. 2 A laboratory setup featuring a glass dish on a stiring hot plate water bath with a three-neck flask containing embryonic hearts being enzymatically digested. Hearts are placed in a spinning flask with 10 mL of ADS buffer, incubated, and stirred in a water bath at 37°C

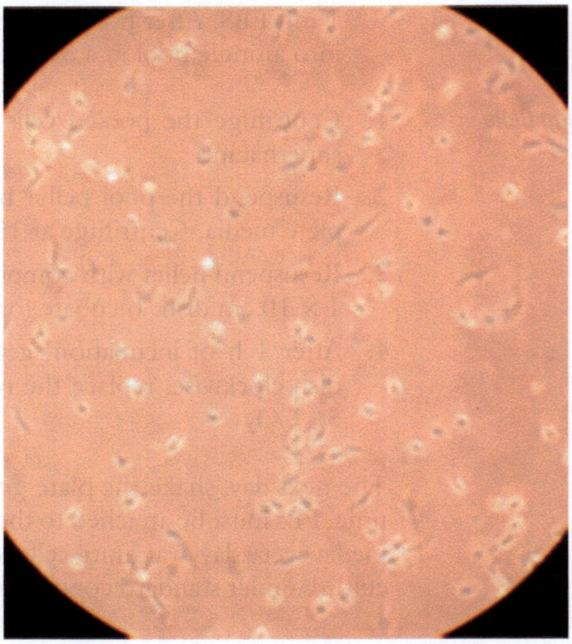

Fig. 3 Microscopic image showing numerous isolated small, round cells cardiac embryonic fibroblast in culture after 7 days of isolation

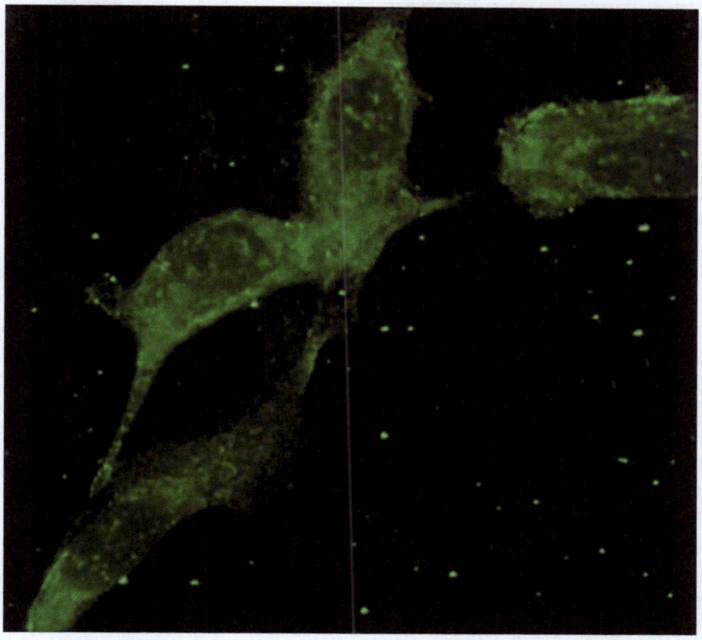

Fig. 4 Fluorescence microscopy image showing green-stained Vimentin for identification of fibroblasts. Spindle-like shape, with a rounded nucleus one of their main morphologic characteristics

3.3 Passage and Freezing Cells

1. When confluent, split cells dividing one 10 cm dish into 3 × 10 cm plates (*see* **Note 5**).
2. To freeze cells, divide a 1 × 10 cm dish into two vials.
3. A few days after the extraction, some fibroblasts were identified by immunofluorescence with a vimentin stain (Fig. 4).

4 Notes

1. These quantities are good enough for a few experiments, as they correspond to more than 100 embryonic hearts. From a single pregnant mouse, 8–10 embryonic hearts can be extracted.
2. Ads buffer without adding the enzymes; therefore, it could be prepared in advance and stored for several days.
3. When collecting the digestion supernatant, you must use the hood at all times to avoid contamination and aspiration of tissue.
4. Digestion must be at 37°C to maintain enzyme activity.
5. When splitting cells, if they are not growing very fast, they can be seeded into a 6 cm dish or 6-well plates, depending on the experiment.

Conflict of Interest

The authors have no conflicts of interest to declare that are relevant to the content of this chapter.

References

1. Tarbit E, Singh I, Peart JN, Rose'Meyer RB (2019) Biomarkers for the identification of cardiac fibroblast and myofibroblast cells. Heart Fail Rev 24(1):1–15
2. Plikus MV, Wang X, Sinha S, Forte E, Thompson SM, Herzog EL, Driskell RR, Rosenthal N, Biernaskie J, Horsley V (2021) Fibroblasts: origins, definitions, and functions in health and disease. Cell 184(15):3852–3872
3. Frangogiannis NG (2019) Cardiac fibrosis: cell biological mechanisms, molecular pathways and therapeutic opportunities. Mol Asp Med 65:70–99
4. Kurose H (2021) Cardiac fibrosis and fibroblasts. Cells 10(7):1716
5. Tallquist MD, Molkentin JD (2017) Redefining the identity of cardiac fibroblasts. Nat Rev Cardiol 14(8):484–491
6. Souders C, Bowers S, Baudino TA (2009) Cardiac fibroblast: the renaissance cell. Circ Res 105(12):1164–1176
7. Lajiness JD, Conway SJ (2012) The dynamic role of cardiac fibroblasts in development and disease. J Cardiovasc Transl Res 5(6):739–748
8. Norris RA, Borg TK, Butcher JT, Baudino TA, Banerjee I, Markwald RR (2008) Neonatal and adult cardiovascular pathophysiological remodeling and repair: developmental role of periostin. Ann N Y Acad Sci 1123:30–40
9. Venugopal H, Hanna A, Humeres C, Frangogiannis NG (2022) Properties and functions of fibroblasts and myofibroblasts in myocardial infarction. Cells 11(9):1386
10. Hall C, Gehmlich K, Denning C, Pavlovic D (2021) Complex relationship between cardiac fibroblasts and cardiomyocytes in health and disease. J Am Heart Assoc 10(5):e019338
11. Fernandes I, Funakoshi S, Hamidzada H, Epelman S, Keller G (2023) Modeling cardiac fibroblast heterogeneity from human pluripotent stem cell-derived epicardial cells. Nat Commun 14(1):8183
12. Gilles G, McCulloch AD, Brakebusch CH, Herum KM (2020) Maintaining resting cardiac fibroblasts in vitro by disrupting mechanotransduction. PLoS One 15(10):e0241390
13. Fan D, Takawale A, Lee J, Kassiri Z (2012) Cardiac fibroblasts, fibrosis and extracellular matrix remodeling in heart disease. Fibrogenesis Tissue Repair 5(1):15
14. Chen W, Bian W, Zhou Y, Zhang J (2021) Cardiac fibroblasts and myocardial regeneration. Front Bioeng Biotechnol 9:599928
15. Cell Applications (1994–2017) Rat cardiac fibroblasts: RCF. https://www.cellapplications.com/rat-cardiac-fibroblasts-rcf. Accessed 11 Feb 2017
16. Almazloum A, Khalil H (2023) Isolation of adult mouse cardiac fibroblasts. Curr Protoc 3(7):e840
17. Sahadevan P, Allen BG (2022) Isolation and culture of adult murine cardiac atrial and ventricular fibroblasts and myofibroblasts. Methods 203:187–195
18. Bekedam FT, Smal R, Smit MC, Aman J, Vonk-Noordegraaf A, Bogaard HJ, Goumans MJ, De Man FS, Llucia-Valldeperas A (2024) Mechanical stimulation of induced pluripotent stem derived cardiac fibroblasts. Sci Rep 14(1):9795
19. Jonsson MKB, Hartman RJG, Ackers-Johnson M, Tan WLW, Lim B, van Veen TAB, Foo RS (2016) A transcriptomic and epigenomic comparison of fetal and adult human cardiac fibroblasts reveals novel key transcription factors in adult cardiac fibroblasts. JACC Basic Transl Sci 1(7):590–602
20. Knight-Schrijver VR, Davaapil H, Bayraktar S, Ross ADB, Kanemaru K, Cranley J, Dabrowska M, Patel M, Polanski K, He X, Vallier L, Teichmann S, Gambardella L, Sinha S (2022) A single-cell comparison of adult and fetal human epicardium defines the ageassociated changes in epicardial activity. Nat Cardiovasc Res 1(12):1215–1229
21. Ferreira R, Hein N (2023) Purification of mouse embryonic fibroblasts (MEFs). Curr Protoc 3(11):e921
22. Hanna M, Sahito RGA, Rateb M, Kachiwal AB, Seddiek HA, Bhutto B, Hescheler J (2020) Generation of transgene-free induced pluripotent stem cells from cardiac fibroblasts of goat embryos. J Stem Cell Regen Med 16(2):34–43
23. Xu J, Wang L, Liu J, Qian L (2021) In vitro conversion of murine fibroblasts into cardiomyocyte-like cells. Methods Mol Biol 2158:155–170

Check for updates

Chapter 11

Standardized Isolation, Culture, and Imaging of Contracting Cardiomyocytes from Neonatal Mouse Hearts

Jielin Yang, Miranda Pannella and Paul Delgado-Olguín

Abstract

Cardiomyocyte maturation represents the final stage of functional development in the heart. During this critical process, fetal cardiomyocytes gradually acquire adult phenotypes and become capable of sustaining billions of contraction–relaxation cycles throughout an organism's lifespan. This transition is driven by gene expression changes that remodel cardiomyocyte structure, metabolism, and functions. In mice, cardiomyocyte maturation takes place during embryonic day (E)18.5 to postnatal two weeks, during which the heart loses its regenerative capacity in response to injury. Hence, characterizing the functional and molecular changes during mouse cardiomyocyte maturation and how this process is regulated provides invaluable insights into heart development and disease. Cardiomyocyte-specific analysis requires isolating cardiomyocytes from other cardiac cell populations. Here, we present a standardized protocol for consistent isolation of neonatal and postnatal mouse cardiomyocytes. Leveraging enzymatic and mechanical digestion, coupled with non-cardiomyocyte depletion via pre-plating, our protocol enables high-quality cardiomyocyte isolation in approximately three hours. In addition, we detail methods for cardiomyocyte culture, molecular, and functional characterization.

Keywords Cardiomyocyte isolation, Neonatal mouse heart, Cardiac maturation, Primary cell culture, Live cell imaging

1 Introduction

The adult mammalian heart is one of the least regenerative tissues. Myocardial injury results in rapid myocyte death that cannot be effectively replenished, leading to pathological remodeling and ultimately heart failure [1]. In contrast, neonatal mouse hearts can effectively regenerate at postnatal day 0 (P0) following severe myocardial injury [2, 3]. However, this regenerative capacity is lost at P7. This temporal pattern coincides cardiomyocyte maturation, a key developmental process that takes place during the perinatal period from embryonic day (E) 18.5 to the first two postnatal weeks [4, 5]. During maturation, cardiomyocytes undergo substantial transcriptional reprogramming that remodels cardiomyocyte structure,

Paul Delgado-Olguín (ed.), *Mouse Embryogenesis*, Methods in Molecular Biology, vol. 3010,
https://doi.org/10.1007/978-1-0716-5126-1_11,

metabolism, and function, enabling the mature heart to sustain billions of cardiac contractions and relaxations over the organism's lifespan [5, 6].

Induced pluripotent stem cell-derived cardiomyocytes (iPSC-CMs) [7] and immortalized cardiomyocyte cell lines (e.g., HL-1) [8] have been invaluable *in vitro* models to study physiological responses to disease conditions and genetic manipulations. However, despite their scalability and ease of genetic modification, these models exhibit key molecular differences to primary cardiomyocytes. For example, iPSC-CMs typically resemble embryonic or early fetal cardiomyocytes and remain structurally, metabolically, and electrophysiologically immature [9]. Therefore, functional and molecular studies of cardiomyocyte maturation require isolating enriched primary cardiomyocyte populations from the developing mouse heart. Understanding the regulatory mechanisms and functional changes underlying cardiomyocyte maturation is crucial for elucidating normal cardiac development and its implications in heart disease and regeneration.

Accurate phenotypic and molecular characterization of cardiomyocyte maturation requires efficient enrichment of cardiomyocytes from other cardiac cell types. Typically, cardiomyocyte isolation involves two sequential steps: heart dissociation and selective enrichment. Static enzymatic digestion is commonly employed for dissociating neonatal mouse hearts but is time intensive [10]. In adult mice, Langendorff retrograde perfusion or ventricular injection are widly adopted methods for digesting the heart [11–13, 16]. Since adult cardiomyocytes are much larger than other cell types, differential gravity precipitation results in cardiomyocyte enrichment at high purity. However, neonatal and early postnatal cardiomyocytes exhibit similar sizes to other cell types in the heart, necessitating alternative enrichment strategies. Magnetic-activated cell sorting (MACS) depletes non-cardiomyocytes with high effectiveness [11]. Nonetheless, the rapid maturation of postnatal cardiomyocytes, characterized by hypertrophic changes, introduces bias in the cell population isolated via MACS columns due to differential size selection. Therefore, a method ensuring consistent isolation of neonatal and postnatal cardiomyocytes is essential for generating comparable data across different stages of cardiomyocyte maturation.

Here, we present a standardized protocol for isolating high-quality cardiomyocytes from neonatal and postnatal mouse hearts, guaranteeing consistent results throughout the various stages of cardiomyocyte maturation. Our protocol combines enzymatic digestion with mechanical agitation to enhance heart dissociation efficiency and leverages the nonattaching characteristic of cardiomyocytes for cell type enrichment. Therefore, we effectively isolate neonatal and postanal cardiomyocytes within a relatively short time frame. Enriched cardiomyocytes are compatible with a wide array of molecular biology assays, including RNA sequencing (RNA-seq), chromatin immunoprecipitation sequencing (ChIP-seq), cleavage under targets and tagmentation (CUT&Tag), and high-throughput chromatin conformation capture (Hi-C). Furthermore, we describe an improved method for culturing synchronized, actively contracting cardiomyocytes.

2 Materials

Reagents are described for a single isolation experiment. One isolation is defined as cardiomyocytes obtained from a pooled set of hearts. Scale accordingly. All reagents, if prepared from salts or lyophilized powders, must be sterile filtered using a 0.22 μm filter before use. All solutions should be sterile and freshly prepared.

2.1 Specimen Preparation

1. Neonatal mouse pups from P0 to P7. A single litter (5–9 pups) is optimal for each isolation. However, a minimum of 1 pup and a maximum of 14 pups can be isolated in a single reaction.
2. 1× Dulbecco's phosphate buffered saline (1× DPBS, no calcium or magnesium). Standard PBS without calcium or magnesium can be used as an alternative.
3. 100 mm and 35 mm sterile Petri dishes.
4. Micro-dissection tools (Dumont #5 forceps and surgical scissors).
5. 70% EtOH.
6. 37°C water bath.

2.2 Heart Tissue Dissociation

1. gentleMACS C Tubes.
2. gentleMACS Octo Dissociator with Heaters.
3. Prepare two enzyme mixes from Neonatal Heart Dissociation Kit (mouse and rat) immediately before harvesting tissues and keep on ice.
   (a) Enzyme Mix 1: add 62.5 μL Enzyme P to 2300 μL Buffer X.
   (b) Enzyme Mix 2: add 25 μL Buffer Y and 12.5 μL Enzyme A to 100 μL Enzyme D.
4. PEB Buffer: Prepare 5 mL of PEB buffer by adding 20 μL of 0.5 M EDTA and 25 μL of 30% BSA to 5 mL of 1× DPBS. Prepare on the day of the experiment and keep on ice. Not required if isolated cells are subsequently cultured.
5. Stop Solution (10% FBS in DMEM): Prepare 15 mL by adding 1.5 mL of fetal bovine serum (FBS) to 13.5 mL DMEM, mix and keep on ice.

6. 1× Red Blood Cell (RBC) Lysis solution: Dilute 10× RBC Lysis solution in ddH_2O.
7. Enzyme A solution: Add 15 μL of Enzyme A from the Neonatal Heart Dissociation Kit to 10 mL of 1× DPBS. Keep on ice.
8. Cell strainer, 70 μm.
9. 50 mL canonical tubes.
10. 1.5 mL microcentrifuge tubes.
11. Trypan blue stain, filter with a 0.22 μm filter before use.
12. Hematocytometer or automated cell counter.

2.3 Cardiomyocyte Enrichment

1. Pre-plating Media (10% FBS in DMEM/F12): prepare one volume (8 mL) per plate used for pre-plating. Calculate total volume needed based on cell input. Prepare an additional volume for washing all the used plates. If a subsequent culture is desired, add Penicillin–Streptomycin to a final concentration of 100 U/mL.
2. 10 cm cell culture dish, tissue culture treated.

2.4 Cardiomyocyte Culture

1. Cardiomyocyte culture medium (all listed with final concentration):
   (a) DMEM/F12 (with 2.5 mM L-glutamine and 1 mM sodium pyruvate; alternatively, manually add L-glutamine and sodium pyruvate)
   (b) 10% FBS
   (c) 100 U/mL Penicillin–Streptomycin
   (d) 1% non-essential amino-acid solution (100×) [for long-term culture]
   (e) 10 mM HEPES [for long-term culture]

   Proliferation inhibitors such as 10 μM AraC (Cytosine β-D-arabinofuranoside hydrochloride) can be added to further enrich cardiomyocytes from proliferating cardiac fibroblasts and endothelial cells during extended culture.
2. Cell culture plates (24-well or 48-well), tissue culture treated. Alternatively, use a glass-bottom chamber slide system or glass cover slips for imaging applications.
3. Laminin solution (10 μg/mL) in DPBS. Avoid freeze-and-thaw cycles.
4. Gelatin solution (0.1% w/v) in DPBS. For 100 mL, prepare by adding 0.1 g of gelatin to 100 mL DPBS. Autoclave to dissolve and sterilize the solution. Store at 4°C.

2.5 Live-Cell Imaging

1. Zeiss Axio Observer 7 epifluorescence microscope, equipped with environmental darkroom chamber (37°C, 5% CO_2).

Alternatively, any microscope capable of capturing time-series images at a frame rate of no less than 10 fps can be used.

2. Cell culture media as described in Sect. 2.4.

3 Methods

All steps except for incubations should be performed at room temperature (RT) (i.e., equilibrate all reagents to RT before adding to cells) to minimize stress due to temperature fluctuations. Fig. 1 illustrates a schematic protocol of major steps involved: heart dissociation (Sects. 3.1 and 3.2), cardiomyocyte enrichment (Sect. 3.3), with cardiomyocyte culture and characterization (Sects. 3.4 and 3.5).

3.1 Heart Tissue Harvesting

1. Prepare a 10 cm Petri dish with room temperature (RT) DPBS for tissue dissection.
2. Use 70% EtOH to sterilize all surgical tools, including tweezers, forceps, and scissors. Alternatively, if isolated cells are used for subsequent cell culture, autoclave the surgical tools and harvest tissue under a biological safety cabinet (BSC).
3. Gently clean the chest of a mouse pup using 70% EtOH.
4. To harvest the heart, perform decapitation to euthanize the neonatal mouse pup.
5. Make an incision along the center of the chest cavity using surgical scissors (Fig. 2) (*see* **Note 1**).
6. Reveal the heart by pressing to open the chest cavity. Use a fine tweezer to grab the base of the heart (near the outflow track) to remove the heart (Fig. 2).
7. Transfer the cardiac tissue into the Petri dish containing DPBS (prepared in step #1). The heart should remain beating to pump out blood remaining in the cardiac chamber.
8. Repeat steps 4–7 for the rest of the pups while keeping the previously excised hearts in DPBS.
9. Separate the ventricles from the atria using tweezers (*see* **Note 2**). Gently open the ventricles and remove any visible blood clots, if any.

3.2 Dissociation of Neonatal Mouse Heart

1. Mix Enzyme Mix 1 and 2. Heat the final enzyme mixture to 37°C for 5 min.
2. During this time, transfer all tissues into a dry, sterile 35 mm Petri dish. Remove excess DPBS transferred along with the tissue (*see* **Note 3**).
3. In the dish, tear the heart tissues into small pieces of size approximately 1 mm^3.

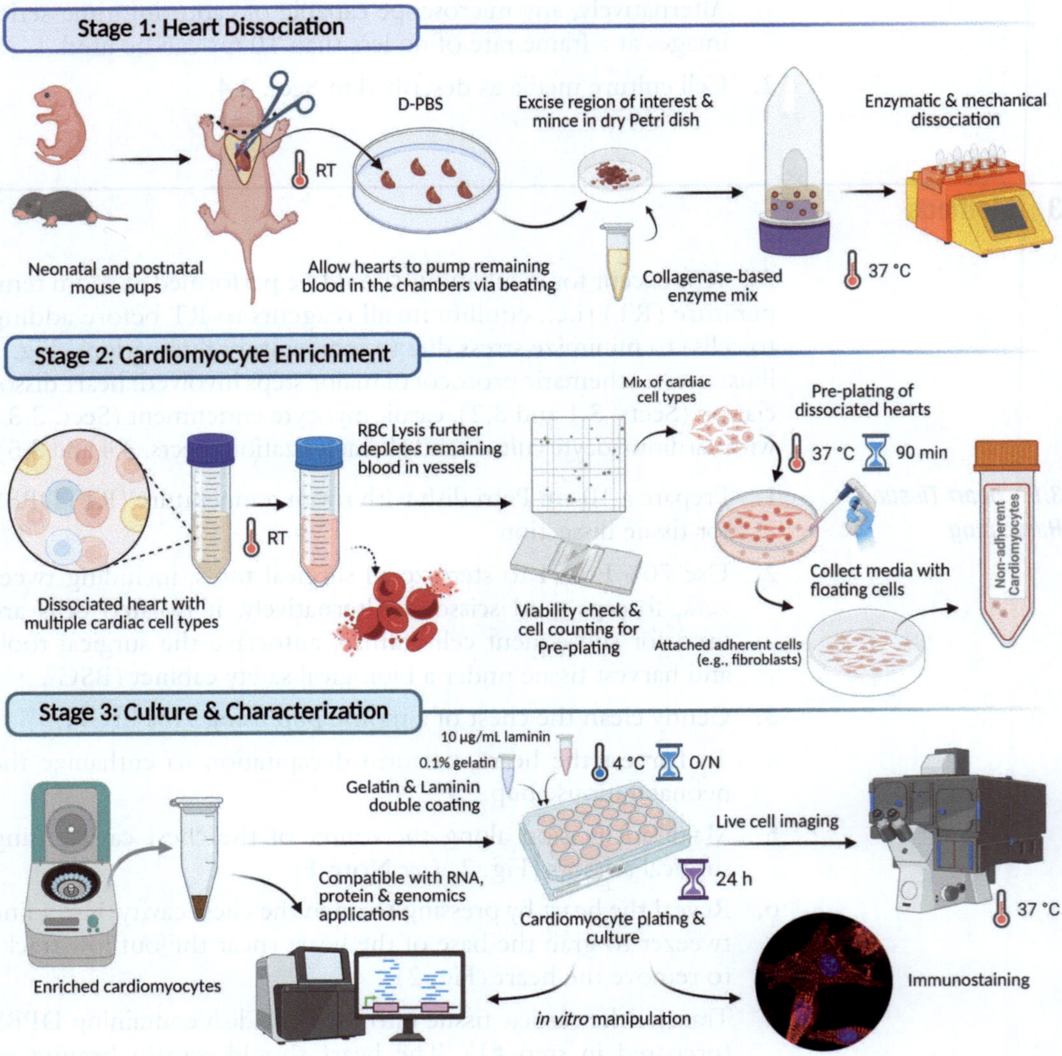

Fig. 1 Protocol schematics of cardiomyocyte isolation from neonatal and postnatal mouse hearts. Figure created with BioRender.com

4. Transfer tissue pieces into a gentleMACS C Tube with tweezers. Add the heated enzyme mix immediately to prevent the tissue from drying out.
5. Gently wash the Petri dish and tweezers with the enzyme mix and transfer all liquid into the C Tube. Secure the cap tightly.
6. Properly attach the C Tube containing the sample to a gentleMACS Octo Dissociator.
7. Run the program 37C_mr_NHDK_1. The duration is approximately 1 hour, during which the heart tissues are dissociated enzymatically and mechanically.

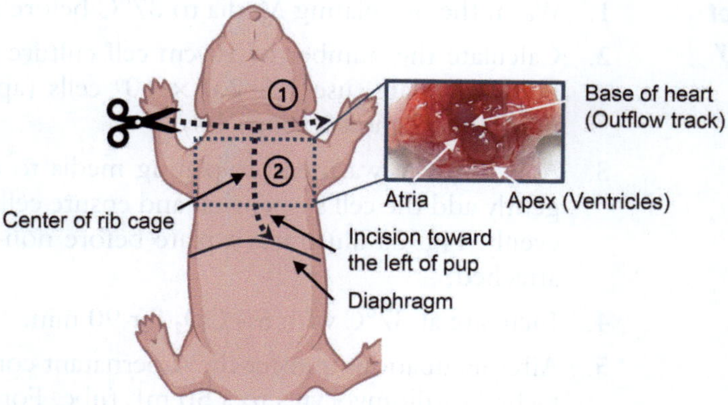

Fig. 2 Procedures for quickly harvesting hearts from neonatal mouse pups. Diagram showing stepwise incisions for (1) euthanizing the pup and (2) opening the chest cavity for optimal exposure of the heart. The incision should not penetrate the diaphragm to prevent moveable organs from the abdomen from blocking the views into the chest cavity when pressure is applied. Applying pressure toward the sides of the chest cavity reveals the heart. The base of the heart can be easily accessed by a fine pair of tweezers to harvest the heart

8. Once the program is terminated, add 7.5 mL Stop Solution and resuspend the cells.
9. Gently apply the cell suspension through the strainer to remove any undigested tissue.
10. Wash the C Tube with 3 mL of Stop Solution and apply through the strainer again. Discard the stainer.
11. Centrifuge the cells at 600 × *g* for 5 min at RT. Aspirate and discard supernatant.
12. Resuspend the cell pellet in 10 mL of RBC Lysis solution. Cells should be resuspended by gentle flicking of the tube.
13. Incubate for 2 min. If RBCs are not visible from the previous cell pallet (step #11), reduce incubation time to 1 min.
14. Immediately add 10 mL of DPBS and flick the tube to mix. PBS/DPBS will stop RBC lysis, and an extended lysis time can damage the integrity of non-RBC cells.
15. Centrifuge at 600 × *g* for 5 min at RT. Aspirate and discard supernatant.
16. Resuspend the cell pellet in 10 mL of Enzyme A solution.
17. Centrifuge at 500 × *g* for 5 min at RT (*see* **Note 4**). Aspirate and discard supernatant.
18. Resuspend the cells in 1–2 mL of pre-plating media (*see* **Note 5**).
19. In a separate 1.5 mL tube, make 1:10 dilution of cells by adding 10 μL of cell suspension and 10 μL of Trypan Blue to 80 μL of PEB Buffer or PBS. Count cells using a hemocytometer or an automated cell counter (*see* **Note 6**).

3.3 Enrichment of Cardiomyocytes by Pre-plating

1. Warm the pre-plating Media to 37°C before proceeding.
2. Calculate the number of 10 cm cell culture plates required. For each plate, use 1.5–2.0 × 10^6 cells (approximately no more than 2 mouse ventricles).
3. Add 8 mL of warmed pre-plating media to each plate, then gently add the cell suspension and ensure cells are distributed evenly. Fig. 3a illustrates a plate before non-cardiomyocytes attached.
4. Incubate at 37°C with 5% CO_2 for 90 min.
5. After incubation, transfer the supernatant containing non-attached cardiomyocytes to a 50 mL tube. For optimal results, each tube should collect no more than 30 mL of media.
6. Gently tap on the bottom of the empty plates 3 times. Wash the plates with 8–10 mL of warm media and combine with the previous cell suspension.
7. Centrifuge at 600×*g* for 5 min at RT. Aspirate and discard supernatant.
8. Resuspend the enriched cardiomyocytes in 2 mL of PEB Buffer or appropriate cardiomyocyte culture media if subsequent cell culture is desired (Sect. 3.4).
9. In a separate 1.5 mL tube, make 1:10 dilution by adding 10 μL of cell suspension and 10 μL of Trypan Blue to 80 μL of PEB Buffer/PBS. Count cells using a hemocytometer or an automated cell counter.
10. Enriched neonatal mouse cardiomyocytes suspension is obtained at this step. Molecular characterizations and other applications can be performed on cell suspension (e.g., RNA/protein extraction). Isolated cardiomyocytes are compatible with other downstream applications where viable, membrane-intact cells are required. The following Sects. 3.4 and 3.5 describe an improved protocol to culture neonatal primary cardiomyocyte that results in beating cardiomyocyte within 24 hours of seeding. Live-cell imaging applications can be used to examine cardiomyocyte contractility (*see* **Note 7**). Representative results are shown in Fig. 3b–f.

3.4 Neonatal Mouse Cardiomyocyte Culture

Here, we describe a protocol that implements serial double-coating of gelatin and laminin for culturing cardiomyocytes. We observe that compared to singly laminin-coated plates, double coating improves cardiomyocyte attachment and synchronized beating within 24 h of seeding.

For steps 1–6 below, perform one day before the isolation experiment (refer to Sects. 3.1--3.3).

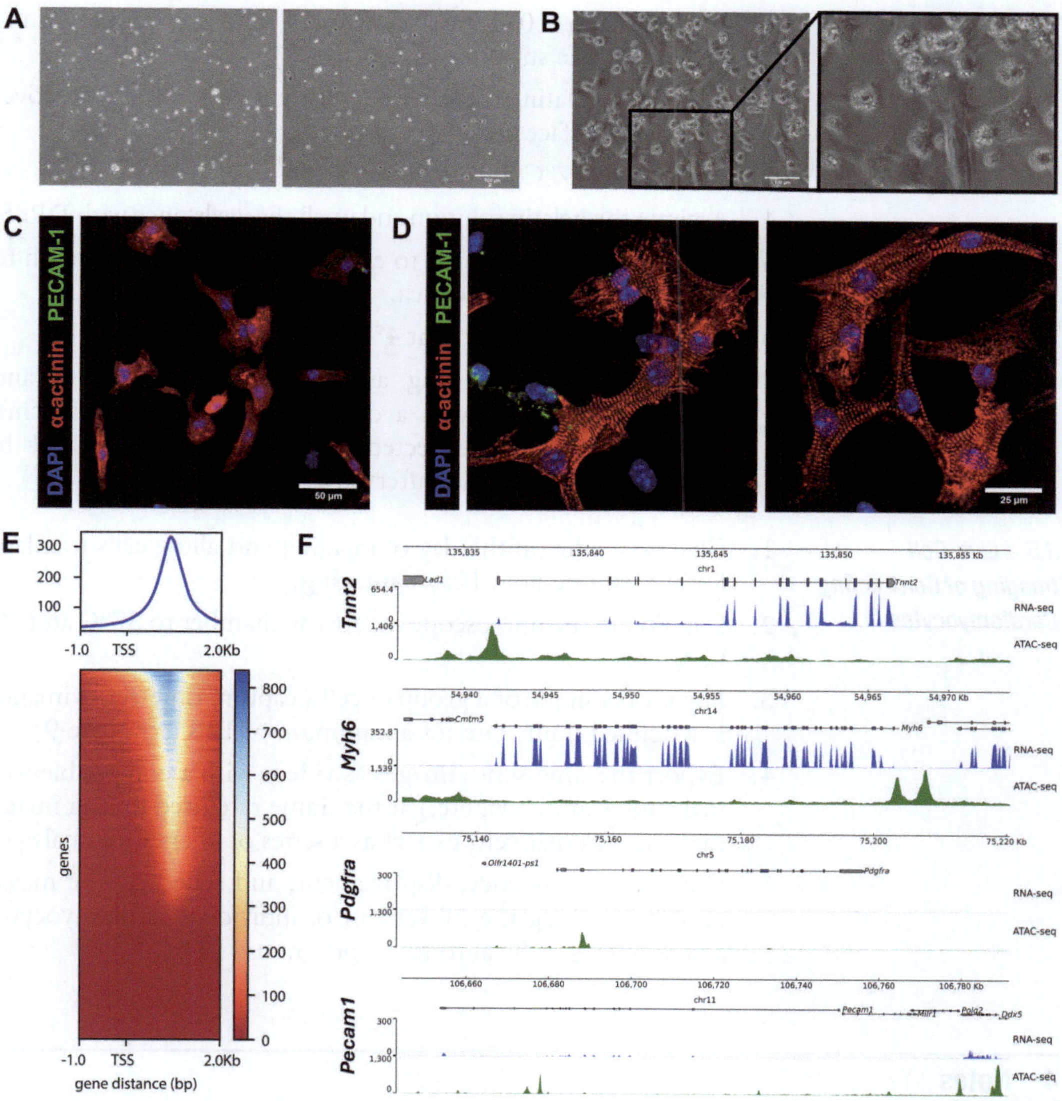

Fig. 3 Characterization of cardiomyocyte during and postenrichment. (**a**) Pre-plating strategy to deplete adherent non-cardiomyocyte populations. The left panel shows dissociated heart cells beginning with pre-plating. The right panel shows adherent non-cardiomyocytes attached to plastic culture ware. (**b**) Brightfield images of cardiomyocytes 48 h after culture. Cardiomyocytes extend and form interconnected myofiber-like structures that allow synchronized beating. (**c**, **d**) Immunostaining of cardiomyocyte and endothelial markers reveals enriched cardiomyocyte populations (**c**) and myofiber-like structures (**d**). (**e**) Profile heatmap of ATAC-seq reads at the transcriptional start site (TSS). (**f**) Representative RNA-seq and ATAC-seq tracks reveal expression and accessible promoters of cardiomyocyte marker genes (*Tnnt2*, *Myh6*) in contrast to fibroblast (*Pdgfra*) and endothelial (*Pecam1*) marker genes

1. Thaw laminin (10 μg/mL) slowly at 4°C to prevent the formation of gel-like structures.
2. Add 0.1% gelatin to each well with enough volume to cover the entire surface area.
3. Incubate at 37°C for 1 h.
4. Aspirate the gelatin solution and wash the wells once with DPBS.
5. Add laminin (10 μg/mL) to each well with volume enough to cover the entire surface area.
6. Coat the plate overnight at 4°C.
7. On the day of cell seeding, aspirate the laminin solution and seed cells immediately at a density of 1.5×10^5 cells/cm^2. Cardiomyocytes are expected to be attached within 24 h. Change media every day afterwards (*see* **Note 8**).

3.5 Live-Cell Imaging of Contracting Cardiomyocytes

1. Change media on the day of imaging and allow cells to adapt for at least one hour before imaging.
2. Equilibrate the microscope incubator chamber to 37°C and 5% CO_2.
3. Focus on a single or a group of cells, capture time-series images at a defined frame rate for a minimum of 12 s (*see* **Note 9**).
4. Export the time-series images as videos with a compatible format (.mp4,.mov,.avi, etc.) at the frame rate used during image capture. Alternatively, export as a series of images for analysis.
5. Contractile frequency, displacement, and speed can be measured by tracking the movement of individual cardiomyocytes across frames or by automated programs [14].

4 Notes

1. Upon cutting through the chest cavity, a small incision toward the left side of the mouse may help reveal the heart.
2. The main protocol described only isolates ventricular cardiomyocytes. Alternatively, if cardiomyocytes of other parts are desired, keep the desired tissues and ensure to remove any blood clots. However, it is important to note that for atrial tissues, it is recommended to dissect and retain only the myocardial region, as connective tissue between the two atrial could affect the efficiency of subsequent tissue digestion and results in a much higher proportion of non-cardiomyocytes in the digested cell population.
3. Reducing carryover PBS ensures optimal enzyme performance at the desired concentration. However, steps where tissue is

not submerged in liquid must be performed as fast as possible to minimize cell death.

4. The final centrifugation is at a lower speed to remove some potential cell debris from the dissociation process.
5. Adjust the volume of media based on the number of hearts dissociated. Scale accordingly. Ensure the concentration is appropriate for cell counting, and for pre-plating, approximately 0.5 mL cell suspension is used per plate.
6. Trypan Blue permeabilizes the dead cell membrane. If viable cells are less than 80%, do not proceed as dead cells can remain in the enriched cardiomyocyte suspension.
7. Isolating cardiomyocytes from P7 mouse hearts will result in lower cardiomyocyte purity compared to using P0 mouse hearts as endothelial cells at later stages shows lower adhesion to TC-treated plastic. Coating plates used for pre-plating with 0.1% gelatin may improve endothelial attachment but may result in slightly lower cardiomyocyte yield.
8. Alternatively, it has been suggested that galactose, instead of glucose, promotes cardiomyocyte maturation *in vitro* [15]. Cell culture media can also be formulated with galactose.
9. Most beating cardiomyocytes often show small displacement. Therefore, a high magnification (20× to 40×) is recommended. If cells are imaged by an epifluorescence microscope, cells should be cultured on a glass-bottom chamber system, since the thickness of plastic plates interferes with high-magnification imaging.

Acknowledgments

This work was supported by the Canadian Institutes of Health Research (CIHR) (162208, 149046, and 468633) and the Heart & Stroke Foundation of Canada (G-25-0040100) to PD-O.

Conflict of Interest The authors have no conflicts of interest to declare that are relevant to the content of this chapter.

References

1. Sutton MG, Sharpe N (2000) Left ventricular remodeling after myocardial infarction: pathophysiology and therapy. Circulation. 101(25):2981–2988
2. Porrello ER, Mahmoud AI, Simpson E, Hill JA, Richardson JA, Olson EN, Sadek HA (2011) Transient regenerative potential of the neonatal mouse heart. Science. 331(6020):1078–1080
3. Porrello ER, Mahmoud AI, Simpson E, Johnson BA, Grinsfelder D, Canseco D, Mammen PP, Rothermel BA, Olson EN, Sadek HA (2013) Regulation of neonatal and adult mammalian

heart regeneration by the miR-15 family. Proc Natl Acad Sci USA 110(1):187–192
4. Kannan S, Kwon C (2020) Regulation of cardiomyocyte maturation during critical perinatal window. J Physiol 598(14):2941–2956
5. Guo Y, Pu WT (2020) Cardiomyocyte maturation: new phase in development. Circ Res 126(8):1086–1106
6. Karbassi E, Fenix A, Marchiano S, Muraoka N, Nakamura K, Yang X, Murry CE (2020) Cardiomyocyte maturation: advances in knowledge and implications for regenerative medicine. Nat Rev Cardiol 17(6):341–359
7. Ahmed A, Syed JN, Chi L, Wang Y, Perez-Romero C, Lee D, Kocaqi E, Caballero A, Yang J, Escalante-Covarrubias Q, Ishimura A, Suzuki T, Aguilar-Arnal L, Gonzales GB, Kim KH, Delgado-Olguín P (2023) KDM8 epigenetically controls cardiac metabolism to prevent initiation of dilated cardiomyopathy. Nat Cardiovasc Res 2(2):174–191
8. White SM, Constantin PE, Claycomb WC (2004) Cardiac physiology at the cellular level: use of cultured HL-1 cardiomyocytes for studies of cardiac muscle cell structure and function. Am J Physiol Heart Circ Physiol 286(3):H823–H829
9. Wu P, Deng G, Sai X, Guo H, Huang H, Zhu P (2021) Maturation strategies and limitations of induced pluripotent stem cell-derived cardiomyocytes. Biosci Rep 41(6):BSR20200833
10. Ehler E, Moore-Morris T, Lange S (2013) Isolation and culture of neonatal mouse cardiomyocytes. J Vis Exp 79:50154
11. Nicks AM, Holman SR, Chan AY, Tsang M, Young PE, Humphreys DT, Naqvi N, Husain A, Li M, Smith NJ, Iismaa SE, Graham RM (2022) Standardised method for cardiomyocyte isolation and purification from individual murine neonatal, infant, and adult hearts. J Mol Cell Cardiol 170:47–59
12. Tian X, Gao M, Li A, Liu B, Jiang W, Qin Y, Gong G (2020) Protocol for isolation of viable adult rat cardiomyocytes with high yield. STAR Protoc 1(2):100045
13. Zhang H, Rabinovitch PS (2022) Protocol for isolation of cardiomyocyte from adult mouse and rat. Bio Protoc 12(10):e4412
14. Ackers-Johnson M, Li PY, Holmes AP, O'Brien SM, Pavlovic D, Foo RS (2016) A simplified, Langendorff-free method for concomitant isolation of viable cardiac myocytes and nonmyocytes from the adult mouse heart. Circ Res 119(8):909–920
15. Maddah M, Heidmann JD, Mandegar MA, Walker CD, Bolouki S, Conklin BR, Loewke KE (2015) A non-invasive platform for functional characterization of stem-cell-derived cardiomyocytes with applications in cardiotoxicity testing. Stem Cell Rep 4(4):621–631
16. Plakhotnik J, Zhang L, Estrada M, Coles JG, Lonnqvist PA, Maynes JT (2022) Local anesthetic cardiac toxicity is mediated by cardiomyocyte calcium dynamics. Anesthesiology 137(6):687–703

Chapter 12

Isolation and Culture of Neonatal Mouse Atrial and Ventricular Myocytes

Yena Oh, Julie Pan and Kyoung-Han Kim

Abstract

Neonatal rodent cardiomyocytes serve as a valuable *in vitro* model for investigating the cellular and molecular mechanisms of the heart, offering an alternative to the technical challenges associated with isolating and maintaining adult cardiomyocytes. Importantly, studying atrial and ventricular cardiomyocytes separately is crucial due to their distinct functional and genetic properties. Here, we present a protocol for the simultaneous isolation and culture of neonatal mouse atrial and ventricular cardiomyocytes. This procedure includes the dissection of neonatal mouse hearts at postnatal day 0 to day 2(P0–P2), separation of atrial appendages and ventricles, and sequential enzymatic digestion using trypsin and collagenase. The chamber-specific identities of isolated cardiomyocytes were validated through gene expression analyses and immunofluorescence staining with atrial and ventricular markers. The resulting cardiomyocyte populations exhibit high viability and retain chamber-specific characteristics, rendering them suitable for a wide array of experimental applications under both physiological and pathological conditions. Potential applications include physiological assessments (e.g., calcium transient analysis and patch clamp), morphological studies, biochemical assays, and drug screening.

Keywords Atria and ventricle, Neonatal, Cardiomyocyte, Cell culture

1 Introduction

Cardiac *in vitro* models are continuously evolving to more accurately recapitulate cardiovascular functions and facilitate the manipulation of molecular processes and cellular behaviors, such as through genetic modifications or changes to environmental conditions (e.g., metabolic energy substrates and hypoxia). Current models, ranging from two-dimensional (e.g., extracellular matrix-patterned cell culture) to three-dimensional (e.g., organoids and engineered heart

Paul Delgado-Olguín (ed.), *Mouse Embryogenesis*, Methods in Molecular Biology, vol. 3010,
https://doi.org/10.1007/978-1-0716-5126-1_12,

tissue) constructs, are able to mimic some of the heart's biological and mechanical properties, including anisotropic tissue structure, extracellular matrix networks, and vascularization [1]. Among the cellular components of the heart, cardiomyocytes, which are the main building blocks responsible for the heart's contractility, constituting the largest portion of its volume, provide invaluable insights into fundamental cardiac functions and mechanisms.

Common cardiomyocyte models include human and rodent-derived immortalized cell lines (e.g., AC16 for ventricular myocytes [2], HL-1 for atrial myocytes [3]), human pluripotent stem cell-derived cardiomyocytes (hPSC-CMs), which include those derived from embryonic stem cells (hESCs-CMs) and induced pluripotent stem cells (hiPSC-CMs), and primary rodent cardiomyocytes. The choice of model depends on factors such as translatability, cellular maturation, production time, cost-effectiveness, genetic manipulation potential, and the ability to distinctly represent atrial and ventricular cardiomyocyte populations.

Studying atrial and ventricular cardiomyocytes separately is essential due to their distinct functions, structures, and genetic profiles [4–6]. Immortalized cell lines allow for independent study of distinct atrial or ventricular populations, but often exhibit diverging phenotypes from primary cells due to the induced genetic mutations required for immortalization. While hPSCs can differentiate into atrial- or ventricular-like cells [7–9], the process is complex and often yields heterogeneous populations. Rodent primary cardiomyocyte isolation protocols have been optimized over the past 50 years, but the isolation of atrial and ventricular cells while achieving high yield and cell viability remains challenging. Protocols for isolating atrial and ventricular cells from adult rat and mouse hearts have been reported [10, 11], but adult rodent cardiomyocytes exhibit lower viability [12] and are prone to losing their phenotype *in vitro* [13, 14]. In contrast, neonatal rodent cardiomyocytes are easier to maintain, offering a more cost-effective alternative for studying cardiac development, physiology and pathology. Despite their immature phenotypes, characterized by differences in metabolism [15] and excitation–contraction coupling processes [16–18], neonatal cardiomyocytes provide valuable mechanistic insights while overcoming many of the limitations of adult cardiomyocytes. Furthermore, neonatal mouse models enable *in vitro* experiments to be performed using the same genetically engineered mouse lines as *in vitro* studies, thereby strengthening the link between experimental systems and enhancing experimental reproducibility and consistency.

Here, we describe a protocol for the simultaneous isolation and culture of neonatal mouse atrial myocytes (NMAMs) and neonatal mouse ventricular myocytes (NMVMs) (Fig. 1). The main steps include the harvest of postnatal day 0-2 (P0–P2) neonatal mouse hearts, microdissection of atrial appendages and ventricles, overnight trypsin predigestion, and subsequent collagenase digestion to release

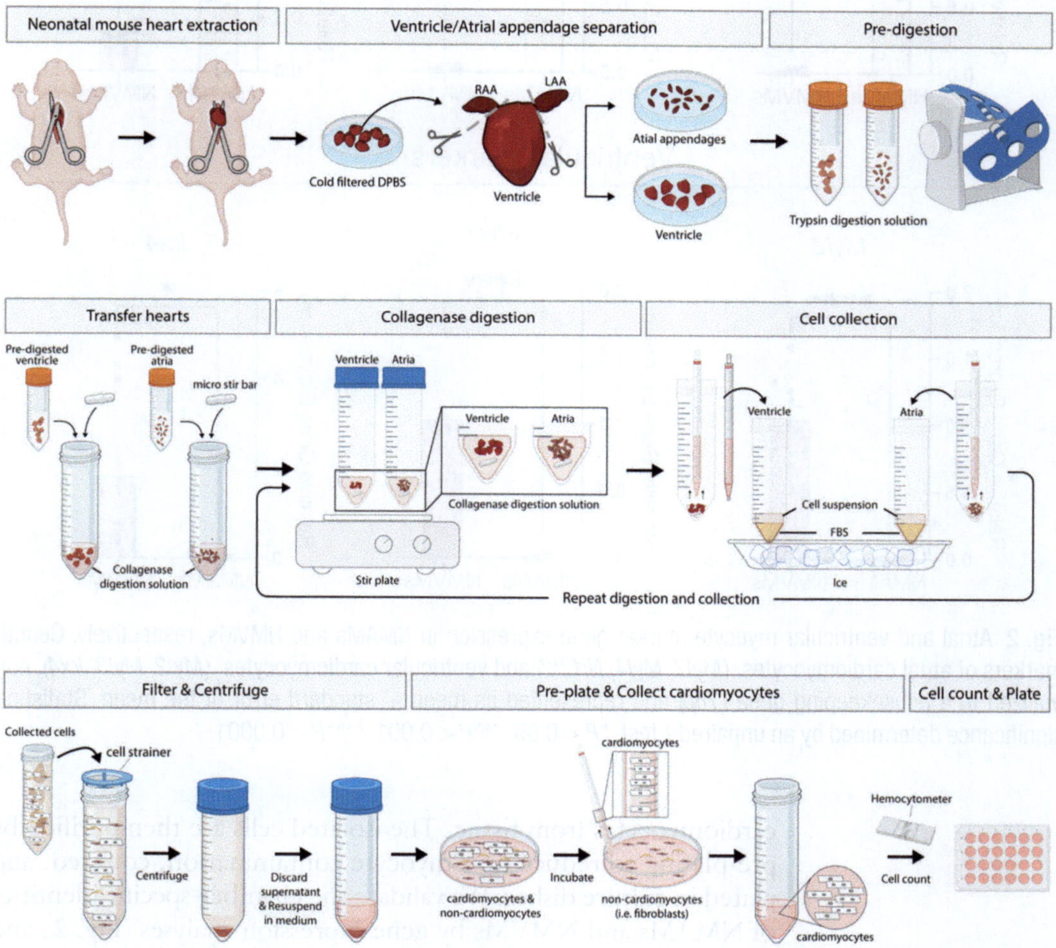

Fig. 1 Schematic of the experimental workflow to isolate neonatal mouse atrial and ventricular cardiomyocytes. Created with BioRender.com

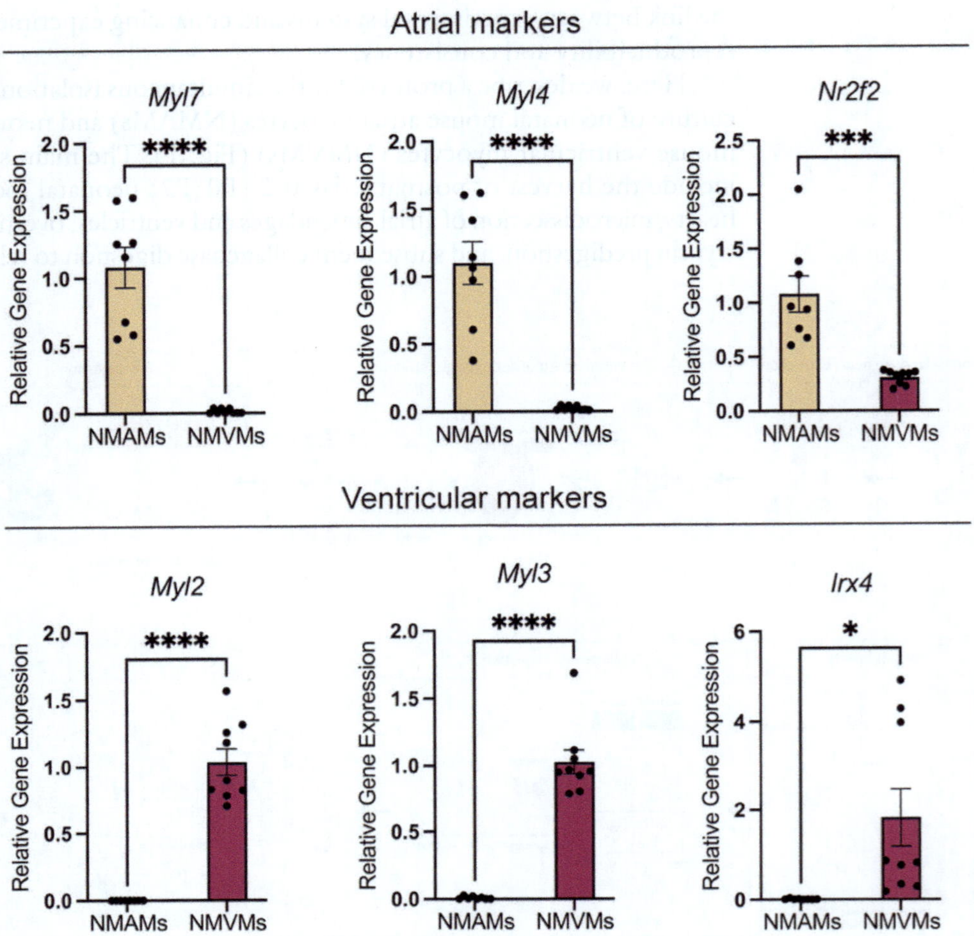

Fig. 2 Atrial and ventricular myocyte marker gene expression in NMAMs and NMVMs, respectively. Genetic markers of atrial cardiomyocytes (*Myl7*, *Myl4*, *Nrf2f2*) and ventricular cardiomyocytes (*Myl2*, *Myl3*, *Irx4*), normalized to a housekeeping gene (*Tbp*) and represented as mean ± standard error of the mean. Statistical significance determined by an unpaired *t*-test. *$P < 0.05$. ***$P < 0.001$. ****$P < 0.0001$

cardiomyocytes from tissue. The isolated cells are then purified by pre-plating to reduce non-myocyte contamination, counted, and plated in culture dishes. We validate the chamber-specific identities of NMAMs and NMVMs by gene expression analyses (Fig. 2) and immunofluorescence staining (Fig. 3) of atrial and ventricular markers. This protocol provides a reliable method for obtaining distinct atrial and ventricular cardiomyocyte populations, and facilitates downstream experiments to elucidate chamber-specific cardiac biology and pathophysiology.

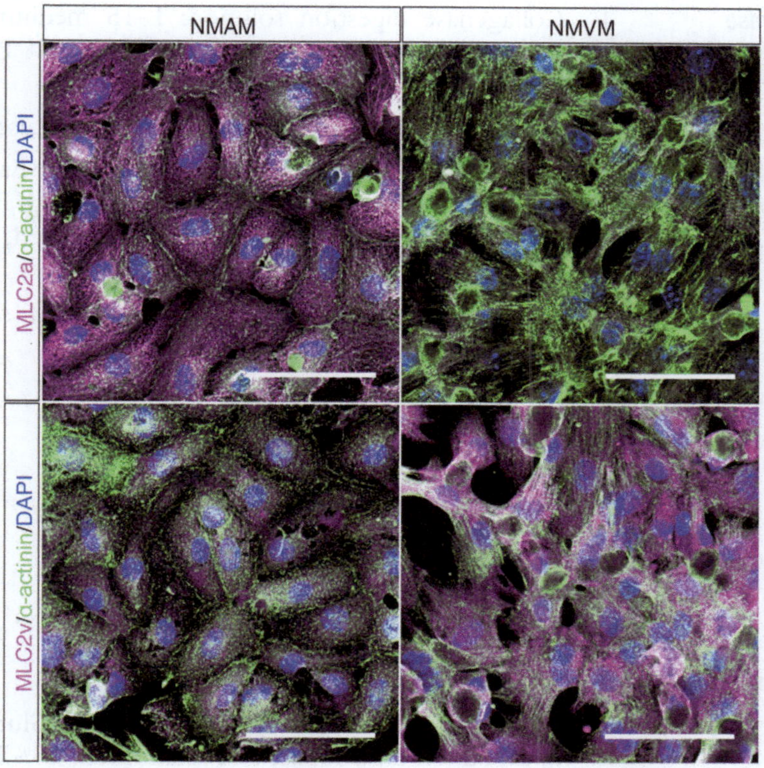

Fig. 3 Immunofluorescent validation of isolated NMAMs and NMVMs. Cells cultured for 96 h on a glass chambered slide were stained for the atrial marker MLC2a (magenta), ventricular marker MLC2v (magenta), pan-cardiomyocyte marker α-actinin (green), and nuclei (DAPI, blue), and imaged by confocal microscopy at 40×. Scale bar = 50 μm

2 Materials

All solutions must be sterile and freshly prepared. All equipment must be sterilized before use.

2.1 Dissection and Predigestion

1. 1× Dulbecco's phosphate-buffered saline (DPBS). Store at 4°C until use.
2. Trypsin predigestion solution: Hank's balanced salt solution (HBSS) with 0.0125% Trypsin–EDTA and 1% Penicillin/Streptomycin (P/S) (*see* **Note 1**).
3. 35 mm cell culture dishes.
4. 70% (v/v) ethanol.
5. Microdissection tools (Dumont No. 5 forceps, scissors, microscissors).
6. Tube rotator or nutator.
7. Digital microscope with screen.

2.2 Collagenase Digestion and Cell Collection

1. Collagenase digestion solution: L-15 medium (Leibovitz) with 0.075% Collagenase II/Dispase and 1% P/S. Incubate at 37°C before use (*see* **Note 2**).
2. Fetal bovine serum (FBS): Store at 4°C before use (*see* **Note 3**).
3. Cell culture medium: Dulbecco's Modified Eagle Medium: Nutrient Mixture F-12 (DMEM/F12) with L-Glutamine, 2.438g/L sodium bicarbonate and phenol red, 10% FBS, 1% P/S. Incubate at 37°C before use.
4. Micro magnetic stir bar (8-10 mm in length) (*see* **Note 4**).
5. Magnetic stirrer.
6. Ice bucket.
7. Serological pipettes (5, 10, and 25 mL).
8. 100 μm cell strainer.
9. Two timers.
10. Centrifuge for cell culture.
11. 37°C bead bath.

2.3 Preplating and Cell Culture

1. Trypan blue solution, 0.4%.
2. Cell culture medium: DMEM/F12 with L-Glutamine, 2.438 g/L sodium bicarbonate and phenol red, 10% FBS, 1% P/S. Incubate at 37°C before use.
3. Cell counter or hemocytometer.
4. For preplating: 35 mm or 100 mm culture dishes (*see* **Note 5**).
5. Incubator (37°C, 5% CO_2, 95% humidity).
6. Inverted microscope.
7. 37°C bead bath.
8. For final plating: Primaria™ Cell Culture Plates
9. Poly-D-Lysine working solution (100 μg/mL)
10. Laminin working solution (15 μg/mL)
11. For immunostaining: Nunc™ Lab-Tek™ Chamber Slide System (8-well)

3 Methods

3.1 Dissection and Tissue Collection

1. Under the biological safety cabinet, prepare 25 mL of predigestion solution in a 50 mL conical tube (*see* **Note 1**).
2. Prepare a minimum of four 35 mm culture dishes, each containing approximately 2 mL of ice-cold DPBS.

3. Transfer one DPBS-filled dish and the neonatal mice to the chemical fume hood.
4. Position the mouse dorsal side up and perform cervical dislocation by applying upward pressure to the posterior cranium while stabilizing the body.
5. Apply 70% (v/v) ethanol to the ventral side of the neonatal mouse for surface sterilization.
6. Reposition the mouse ventral side up. Using fine forceps, retract the forelimbs to expose the thoracic cavity.
7. Make an incision along the right lateral side of the sternum. Using slightly open dissection scissors, apply gentle downward pressure to expose the heart, and cut underneath it to remove the heart.
8. Transfer the collected heart to the first DPBS-filled dish.
9. Repeat steps 4–8 for each neonatal mouse.
10. Once all hearts are collected, remove any noncardiac tissues (e.g., lung), and transfer hearts to the second dish containing fresh cold DPBS to remove excess blood.
11. Microdissect the atrial appendages from the ventricles under a digital microscope with a screen or a magnifier. Place atria in the third dish, and ventricles in the last dish.
12. Make 5–6 incisions in each ventricle while keeping the tissue intact (*see* **Note 5**).
13. Using a 10 mL pipette, carefully aspirate the atria and ventricles from each dish and transfer them to separate 50 mL conical tubes, each containing 25 mL of the trypsin predigestion solution (*see* **Note 6**).
14. Place the tubes onto a tube rotator or nutator in the 4°C cold room and rotate for 14–16 h at approximately 12 RPM (*see* **Note 7**).

3.2 Collagenase Digestion

Perform the following steps for both the atria and ventricle, unless otherwise stated:

1. Transfer the predigestion tubes to the biological safety cabinet.
2. Allow tissue to settle at the bottom of the tube, then carefully remove the trypsin predigestion solution using a 25 mL pipette. Avoid overdisrupting tissue (*see* **Note 8**).
3. Add 3 mL of prewarmed (37°C) collagenase digestion solution to each of the atrial and ventricular tissue tubes.
4. Add a sterile micro stir bar to each of the tubes and place the tubes in a tube rack above the magnetic stirrer. Set the stirrer to 50–100 RPM.

5. Stir the ventricular tissue for 15 min and the atrial tissue for 5 min at room temperature (RT).
6. Using a 10 mL pipette, gently aspirate and release the tissue up and down the pipette 3–4 times and wait briefly for 5–10 s to allow the tissue to settle to the bottom of the tube.
7. Carefully collect the supernatant containing isolated cells and transfer it to a 50 mL tube containing 5 mL of cold FBS (4°C) (*see* **Note 3**). Place the tube on ice.
8. Repeat steps 3–7 (*see* **Note 9**), collecting the digested solution in separate FBS tubes for atria and ventricles.

3.3 Preplating and Cardiomyocyte Collection

Perform the following steps for both the atria and ventricle, unless otherwise stated:

1. Strain the collected cells (in FBS) into two new 50 mL conical tubes through a 100 μm cell strainer using a 25 mL pipette.
2. Balance and centrifuge the tubes at 200 × *g* (approximately 1,000 rpm) for 5 min at RT (*see* **Note 10**).
3. Gently pour out the supernatant into a waste bottle by tilting the 50 mL tube (*see* **Note 11**).
4. Add 2 mL of media (DMEM/F12 + 10% FBS + 1% P/S) to each cell pellet. Thoroughly but gently resuspend the pellet using a P1000 pipette (*see* **Note 12**). Avoid making bubbles.
5. Prepare a 35 mm culture dish for atrial cardiomyocytes and a 100 mm culture dish for ventricular cardiomyocytes (*see* **Note 13**). Add 8 mL of media to the 100 mm dish. Gently resuspend cells again and transfer the 2 mL of cells from each tube into each dish (*see* **Note 14**).

3.4 Cardiomyocyte Isolation and Culture

1. Preplate the cells for 1 h in an incubator (37°C, 5% CO_2).
2. Remove the culture dish from the incubator and gently pipette the supernatant (containing unattached cardiomyocytes) up and down to collect weakly attached cardiomyocytes. Transfer the cardiomyocyte-containing supernatant using a 10 mL pipette to a new 50 mL Falcon tube (*see* **Note 15**). Avoid making bubbles.
3. Count the number of cells using a hemocytometer and Trypan blue stain (*see* **Note 16**).
4. Plate the cardiomyocytes in a culture plate at the appropriate density for downstream experiments (*see* **Note 17**).
5. Incubate overnight (37°C, 5% CO_2).
6. Assess cell viability and morphology the following day under a microscope.

7. Wash the cells with prewarmed (37°C) culture medium or DPBS.
8. Proceed with experimental protocols (*see* **Note 18**).

3.5 (Optional) Double-Coating of Glass Slide with Poly-D-Lysine and Laminin

For immunostaining purposes, we utilize the following double-coating protocol adapted from Pereira et al. [19] and culture the cells in a double-coated 8-well chambered glass slide.

1. Prepare a 100 μg/mL Poly-D-Lysine working solution in sterile distilled water.
2. Coat the bottom of each well with the working solution.
3. Incubate the coated plates at RT for 1–2 h.
4. Aspirate the Poly-D-Lysine solution and rinse the wells three times with distilled water (*see* **Note 19**). Wait until the slide is air-dried before proceeding with the next step.
5. Prepare a 15 μg/mL laminin working solution in sterile distilled water.
6. Coat the bottom of each well with the working solution.
7. Incubate the coated slides at 37°C for 2 h (or 4°C overnight).
8. Bring the coated slide to RT before aspirating and discarding the laminin solution (*see* **Note 20**).
9. Immediately plate the cells on the coated slide (*see* **Note 17**).

4 Notes

1. Prepare 25 mL of predigestion solution per tube, accommodating approximately 40 atria or 20 ventricles per tube.
2. More than 10 rounds of collagenase digestion (3 mL per round) will be required for complete tissue digestion.
3. FBS neutralizes the enzymatic activity of collagenase, preventing excessive digestion of collected cells and promoting cell survival.
4. For sterilization, place micro stir bars in a 35 mm culture dish containing 70% EtOH using forceps. After 1–2 min of incubation, remove EtOH and air-dry the stir bars.
5. Cutting ventricles open promotes efficient digestion by increasing the surface area exposed to digestion enzymes. Avoid mincing the tissue; ventricles should remain largely intact after cutting. Overcutting the ventricles may lead to overdigestion. Make a transverse cut at least once to allow blood inside the chambers to wash out and enhance digestion.

6. Heart tissues are highly adhesive; excessive pipetting can cause the heart to stick to the inner walls of the serological pipette, resulting in tissue loss. Ensure the heart stays near the bottom of the pipette during handling.
7. Properly balance the tube rotator or nutator to ensure consistent agitation during overnight digestion.
8. Clumping of predigested tissues is normal. Handle the tissues gently to avoid damage during pipetting.
9. Complete tissue digestion is characterized by the loss of structural integrity and a viscous consistency. Adjust collagenase digestion time and increase collection frequency to prevent overdigestion.
10. Equilibrate FBS volumes between tubes if necessary. Post-centrifugation, cells will form a dark pink pellet at the tube bottom.
11. Do not tilt the 50 mL tube too slowly when pouring out the supernatant, as it may result in cell loss.
12. Thoroughly but gently resuspend the cell pellet to ensure a homogeneous single-cell suspension. This will be crucial for uniform plating and to prevent cellular aggregation.
13. Adjust the number of pre-plating culture vessels and its dimensions, media volume, and cell density based on the number of isolated hearts and experimental requirements. For example, when processing approximately 20 neonatal mouse hearts, one 100 mm culture dish containing 8 mL of media is typically sufficient for ventricular cardiomyocytes. For less abundant atrial cardiomyocytes, use smaller vessels (e.g., a 35 mm dish with 2 mL media) to optimize cell concentration. Consider downstream applications (e.g., imaging and pharmacological treatments) when determining pre-plating conditions.
14. Check cell plating under the microscope to ensure even distribution. Cells will initially appear round and clear when first plated.
15. Fibroblasts, being highly adhesive, will stick to the plate during preplating. Cardiomyocytes remain in suspension or weakly adherent, allowing for the selective removal of fibroblasts [20]. After pre-plating and the removal of the supernatant containing cardiomyocytes, consider gently rinsing the pre-plated dish with fresh media to recover any remaining cardiomyocytes.
16. Efficient digestion typically yields ~0.5×10^6 cells per neonatal mouse ventricle.

17. For RNA isolation, seeding approximately 2–4 × 10^5 cardiomyocytes per well in a 24-well culture plate (1.9 cm^2 per well) typically yields 1–3 μg of RNA using Trizol extraction. Figure 2 shows quantitative PCR (qPCR) performed on NMAMs and NMVMs seeded at these densities, validating atrial (*Myl7*, *Myl4*, *Nr2f2*) and ventricular (*Myl2*, *Myl3*, *Irx4*) marker gene enrichment in their respective subpopulations. For imaging purposes, seeding 1 × 10^5 cells in an 8-well chambered slide (0.8 cm^2 per well) achieves ~70–80% confluency, suitable for high-density cell visualization. Figure 3 shows the immunofluorescence staining of NMAMs and NMVMs seeded at this density, further verifying their chamber identities with enrichment of *Myl7* (MLC2a) and *Myl2* (MLC2v). When assessing cellular hypertrophy, plating 2 × 10^5 cells per well in a 24-well plate achieves ~50% confluency, which helps delineate cell size as ventricular cardiomyocytes tend to cluster at higher densities.
18. Under the microscope, verify cell health by examining morphology (e.g., expanded and noncircular shape indicating proper adhesion). Check for signs of contamination or other abnormalities, such as blebbing or excessive granularity. If cells appear suboptimal, they can be cultured for an additional 24 h.
19. Ensure that the poly-D-lysine is thoroughly rinsed, as residual amounts can be toxic to cells [21].
20. Avoid drying the coated wells after laminin application, as drying disrupts laminin's structure and adhesive properties. Washing after laminin coating is unnecessary.

Acknowledgments

This work was supported by the Natural Sciences and Engineering Research Council of Canada Discovery Grant (RGPIN-2025-05100) and partly by the Canadian Institutes of Health Research Project Grants (PJT 173281 and PJT 195676) and Heart and Stroke Foundation in Canada Grant-in-Aid (G-25-0041680) to Dr. Kyoung-Han Kim. Yena Oh was supported by the Frederick Banting and Charles Best Canada Graduate Scholarships Doctoral Award. Julie Pan was supported by the Canada Graduate Scholarship-Master's Program.

Conflict of Interest The authors have no conflicts of interest to declare that are relevant to the content of this chapter.

References

1. Jorba I, Mostert D, Hermans LHL, van der Pol A, Kurniawan NA, Bouten CVC (2021) In vitro methods to model cardiac mechanobiology in health and disease. Tissue Eng Part C Methods 27(3):139–151
2. Davidson MM, Nesti C, Palenzuela L, Walker WF, Hernandez E, Protas L, Hirano M, Isaac ND (2005) Novel cell lines derived from adult human ventricular cardiomyocytes. J Mol Cell Cardiol 39(1):133–147
3. Claycomb WC, Lanson JrNA, Stallworth BS, Egeland DB, Delcarpio JB, Bahinski A, Izzo JrNJ (1998) HL-1 cells: a cardiac muscle cell line that contracts and retains phenotypic characteristics of the adult cardiomyocyte. Proc Natl Acad Sci USA 95(6):2979–2984
4. Oh Y, Abid R, Dababneh S, Bakr M, Aslani T, Cook DP, Vanderhyden BC, Park JG, Munshi NV, Hui CC, Kim KH (2024) Transcriptional regulation of the postnatal cardiac conduction system heterogeneity. Nat Commun 15(1):6550
5. urashnikov A, Di Diego JM, Zygmunt AC, Belardinelli L, Antzelevitch C (2007) Atrium-selective sodium channel block as a strategy for suppression of atrial fibrillation: differences in sodium channel inactivation between atria and ventricles and the role of ranolazine. Circulation 116(13):1449–1457
6. Wu SP, Cheng CM, Lanz RB, Wang T, Respress JL, Ather S, Chen W, Tsai SJ, Wehrens XH, Tsai MJ, Tsai SY (2013) Atrial identity is determined by a COUP-TFII regulatory network. Dev Cell 25(4):417–426
7. Lee JH, Protze SI, Laksman Z, Backx PH, Keller GM (2017) Human pluripotent stem cell-derived atrial and ventricular cardiomyocytes develop from distinct mesoderm populations. Cell Stem Cell 21(2):179–194.e4
8. Devalla HD, Schwach V, Ford JW, Milnes JT, El-Haou S, Jackson C, Gkatzis K, Elliott DA, Chuva, de Sousa Lopes SM, Mummery CL, Verkerk AO, Passier R (2015) Atrial-like cardiomyocytes from human pluripotent stem cells are a robust preclinical model for assessing atrial-selective pharmacology. EMBO Mol Med 7(4):394–410
9. Bardot E, Calderon D, Santoriello F, Han S, Cheung K, Jadhav B, Burtscher I, Artap S, Jain R, Epstein J, Lickert H, Gouon-Evans V, Sharp AJ, Dubois NC (2017) Foxa2 identifies a cardiac progenitor population with ventricular differentiation potential. Nat Commun 8: 14428
10. Plačkić J, Kockskämper J (2018) Isolation of atrial and ventricular cardiomyocytes for in vitro studies. Methods Mol Biol 1816:39–54
11. Blackwood EA, Bilal AS, Azizi K, Sarakki A, Glembotski CC (2020) Simultaneous isolation and culture of atrial myocytes, ventricular myocytes, and non-myocytes from an adult mouse heart. J Vis Exp (160). https://doi.org/10.3791/61224
12. Banyasz T, Lozinskiy I, Payne CE, Edelmann S, Norton B, Chen B, Chen-Izu Y, Izu LT, Balke CW (2008) Transformation of adult rat cardiac myocytes in primary culture. Exp Physiol 93(3):370–382
13. Kubin T, Pöling J, Kostin S, Gajawada P, Hein S, Rees W, Wietelmann A, Tanaka M, Lörchner H, Schimanski S, Szibor M, Warnecke H, Braun T (2011) Oncostatin M is a major mediator of cardiomyocyte dedifferentiation and remodeling. Cell Stem Cell 9(5):420–432
14. Ellingsen O, Davidoff AJ, Prasad SK, Berger HJ, Springhorn JP, Marsh JD, Kelly RA, Smith TW (1993) Adult rat ventricular myocytes cultured in defined medium: phenotype and electromechanical function. Am J Physiol 265(2 Pt 2):H747–754
15. Lopaschuk GD, Spafford MA, Marsh DR (1991) Glycolysis is predominant source of myocardial ATP production immediately after birth. Am J Physiol 261(6 Pt 2):H1698–H1705
16. Diebold RJ, Koch WJ, Ellinor PT, Wang JJ, Muthuchamy M, Wieczorek DF, Schwartz A (1992) Mutually exclusive exon splicing of the cardiac calcium channel alpha 1 subunit gene generates developmentally regulated isoforms in the rat heart. Proc Natl Acad Sci USA 89(4):1497–1501
17. Page E, Buecker JL (1981) Development of dyadic junctional complexes between sarcoplasmic reticulum and plasmalemma in rabbit left ventricular myocardial cells. Morphometr Anal Circ Res 48(4):519–522
18. Sedarat F, Xu L, Moore ED, Tibbits GF (2000) Colocalization of dihydropyridine and ryanodine receptors in neonate rabbit heart using confocal microscopy. Am J Physiol Heart Circ Physiol 279(1):H202–H209
19. Pereira AHM, Cardoso AC, Franchini KG (2021) Isolation, culture, and immunostaining of neonatal rat ventricular myocytes. STAR Protoc 2(4):100950
20. Richler C, Yaffe D (1970) The in vitro cultivation and differentiation capacities of myogenic cell lines. Dev Biol 23(1):1–22
21. Faussner A, Deininger MM, Weber C, Steffens S (2022) Direct addition of poly-lysine or poly-ethylenimine to the medium: a simple alternative to plate pre-coating. PLoS One 17(7):e0260173

Check for updates

Chapter 13

MicroRNA Quantification on Fast and Slow-Twitch Skeletal Muscles of Mouse

Magally Ramírez-Ramírez, María Luisa Benítez-Hess, Estela G. García-González and J. Manuel Hernández-Hernández

Abstract

microRNAs (miRNAs) are noncoding RNAs of approximately 24 nucleotides long that are expressed in multiple tissues and organs. miRNAs regulate the expression of specific target genes, and their expression profile can provide insight into the cell biology altered by diseases or drug treatments. Research has demonstrated that several miRNAs regulate muscle differentiation and the specification of fiber type. Hence, an established tool to routinely quantify miRNA abundance is of the utmost importance. The stem–loop quantitative reverse transcription PCR (RT-qPCR) is a methodology that allows the analysis of a specific miRNA with high sensitivity. Upon hybridization with a structured stem-loop oligonucleotide to the 3′ end of the target miRNA, a retro-transcription is possible for further qPCR quantification. In this chapter, we describe a protocol to quantify miRNAs in dissected mouse fast and slow-twitch muscle.

Keywords MicroRNA, Fast-twitch fiber, Slow-twitch fiber, Skeletal muscle, Stem–loop PCR, RNA isolation, miRNAs isolation

1 Introduction

In recent years, several studies have been conducted regarding the participation of miRNAs in different processes and pathologies directly affecting skeletal muscle biology [1–3]. miRNAs are small, single-strand noncoding RNAs that participate in gene expression regulation [4]. To correlate the expression of target genes and explain specific phenotypes it is necessary to know the differential miRNAs abundance.

The stem–loop quantitative reverse transcription PCR (RT-qPCR) allows the analysis of multiple miRNAs. First, the stem–loop retro-transcription (RT) primer is hybridized to a specific miRNA molecule and reverse-transcribed in an RT reaction. Then, the RT product is amplified using a miRNA-specific forward oligonucleotide and a universal reverse primer. Due to steric hindrance,

Paul Delgado-Olguín (ed.), *Mouse Embryogenesis*, Methods in Molecular Biology, vol. 3010,
https://doi.org/10.1007/978-1-0716-5126-1_13,

it is important to note that the stem–loop oligonucleotide structure does not allow the hybridization with the pri- or pre-miRNA. Lastly, the generated cDNA is quantified by qPCR using SYBR Green detection technology.

This protocol describes the correct recognition and dissection of a mouse tibialis anterior and diaphragm muscles, the isolation of total RNA and the enrichment of miRNAs. Additionally, it also shows how to structure stem loops and quantitative analysis of example miRNAs by qPCR.

2 Materials

Prepare all solutions using distilled ultra-pure water. Store all reagents at 4°C or −20°C, unless indicated otherwise. Treat waste materials following international regulations. All reagents used are molecular biology grade.

2.1 Buffers and Solutions

1. Nucleic acid loading sample buffer: 1× (50 mM Tris-HCL, pH 8.0, 0.25% Glycerol, 5 mM EDTA, 0.2% Bromophenolblue, 0.2% Xylene Cyanole FF Blue. Store at room temperature.
2. TBE 1× (Tris, boric acid and EDTA). Electrophoresis running buffer. 89 mM Tris, 89 mM boric acid, 2 mM EDTA, pH 8.3. Store at room temperature.
3. Agarose 1.5% in TBE 1×. Freshly prepared.

2.2 Laboratory Equipment

1. Micropipettes and filtered pipette tips (2, 10, 200, and 1000 μL).
2. Polypropylene thin-wall 0.2 μL micro centrifuge tubes. RNase free.
3. Polypropylene 0.5 mL and 1.5 mL microcentrifuge tubes. RNase-free tubes.
4. Microcentrifuges (BioScience).
5. Heat blocks (Lab net).
6. Vortex Genie 2 (Scientific Industries).
7. PCR machine (Axygen).
8. Power supply Power PAC 3000 (Bio Rad).
9. Real-time qPCR machine Rotor-Gene ® RG-3000 real-time thermocycler (Qiagen) with excitation (470 nm) and emission (510 nm) filters for the detection of SYBR Green.
10. Photo documentation system for agarose gels.
11. Nanodrop Nucleic acids quantification equipment.
12. Midi Horizontal electrophoresis apparatus system (Cleaver Scientific).

3 Methods

3.1 Tibialis Anterior and Diaphragm Dissection

1. For each muscle, have two 60 mm culture dishes with enough cold sterile PBS, and keep them on ice.
2. Euthanize mice following institutional-approved procedures.
3. Before dissection of the tissue of interest, spray the abdominal area and the lower limbs of the euthanized animal with 70% ethanol.
4. Place the mouse on its back in a supine position, leaving the lower limbs accessible for dissection. To separate the fur from the muscles, use a scalpel, scissors, and tweezers. Locate and expose the *tibialis anterior* muscle and make three cuts to release the complete muscle. The first cut allows the release of the fascia surrounding the *TA*. Another cut is performed in the ankle tendon, and the third in the upper part of the palletar tendon.
5. Dissect the Diaphragm after the chest is opened and the rib cage is exposed. Holding the sternum with forceps, a cut in the Diaphragm is made surrounding the rib cage.
6. Place the dissected muscles on culture dishes containing cold PBS to gently remove the excess blood, adipose tissue, and tendons (*see* **Note 2**).
7. Remove the excess PBS with a Kim wipe and transfer each muscle to a new 1.5 mL tube containing RNA later.
8. Store the samples at 4°C for further RNA purification.

3.2 microRNA Extraction from Skeletal Muscle

1. Cut the isolated muscle into small pieces and transfer it to a 1.5 mL tube with 1 mL of Trizol.
2. Homogenize the tissue using a manual tissue pestle grinder. Vortex the sample for 30 s and incubate 5 min at 4°C.
3. Add 200 μl of chloroform to the sample and vortex for 10 s.
4. Centrifuge for 10 min at 12,000 rpm at 4°C (*see* **Note 2**).
5. Collect the upper phase containing the nucleic acids (approximately 540 μL) and transfer it to a new 1.5 mL microtube.
6. Add 1.5 volumes of absolute ethanol and mix by pipetting up and down the solution.
7. Place an EZ-10 Spin Column in a 2.0 mL Collection tube and transfer 750 μL of the mixture to the column. Spin at 12,000 rpm for 2 min at 4°C. Discard the flow-through and place the column back in the same collection tube. Reload the column with RNA–ethanol mix, centrifuge for 30 s at the same speed and temperature, as in step 7.

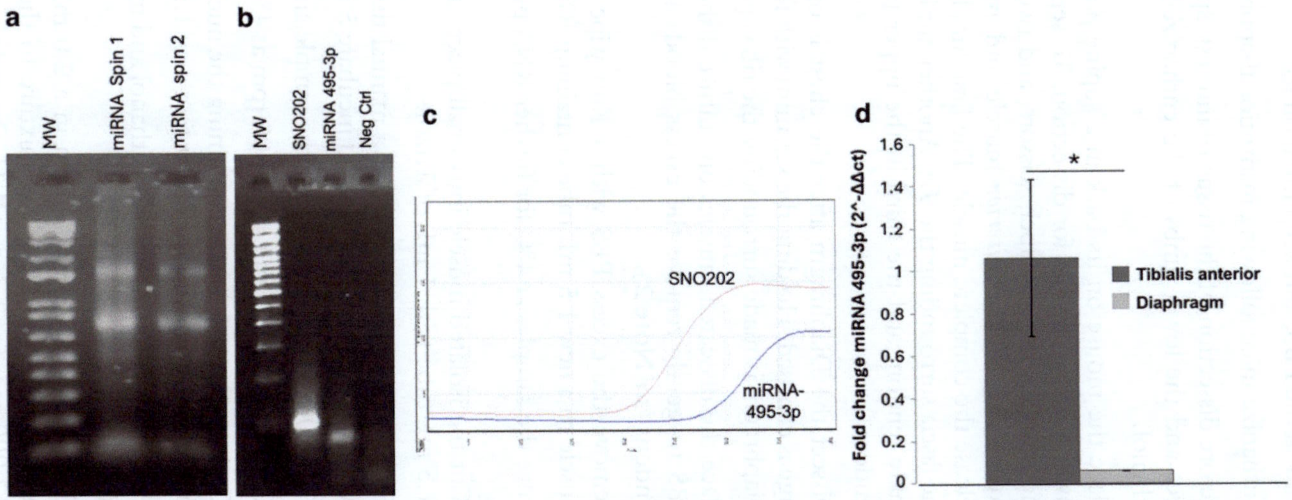

Fig. 1 Representative results of miRNAs isolation and quantification. (**a**) Total RNA integrity on a 1.5% agarose gel shows the presence of small RNAs. (**b**) End-point PCR analysis of positive control SNO202 and miRNA-495-3p shows specific fragments of 100 and 66 bp, respectively. (**c**) Representative amplification curve of SNO202 (red) and miRNA 495-5p (red). (**d**) Differential expression of miRNA-495 in Tibialis anterior and Diaphragm.

Table 1
Oligonucleotide sequences

miR495-3p	Stem–loop 5′-GTCGTATCCAGTGCAGGGTCCGAGGTATTCGCACTGGAT ACGACAAGAAG-3′ Forward oligonucleotide 5′-GTCGTATCCAGTGCAGGGTCCGAGGTATTCGCACTGGAT ACGACAAGAAG-3′
SNO 202	Stem–loop 5′-GTCGTATCCAGTGCAGGGTCCAAAATATGGAACGCT TCACGAATT TG-3′ Forward oligonucleotide 5′-CTCCTTCGGCAGCACATATACT-3′
Universal Reverse	5′-GTG CAG GGT CCG AGG T-3′

8. Add 500 μL of RPE buffer to the center of the column, centrifuge at 12,000 rpm for 30 s at 4°C. Discard the supernatant and wash once more.
9. Place the column in a new microcentrifuge tube and add 40 μL of ultra-pure water at the center of the membrane within the column. Incubate 2 min at room temperature (*see* **Note 3**).
10. Centrifuge at 12,000 rpm for 30 s at 4°C.
11. Discard the column and keep the purified RNA samples at −70°C for later use (*see* **Note 4**).
12. Confirm the miRNA integrity by running a 1.5% denaturing agarose gel stained with ethidium bromide (Fig. 1a).

3.3 Oligonucleotide Design

1. Design stem–loops RT oligonucleotides combining a universal backbone sequence of 44 nucleotides, designed to form a stem–loop structure. This sequence is: 5′-GTC GTA TCC AGT GCA GGG TCC GAG GTA TTC GCA CTG GAT ACG AC-3′. Add a complement of six 3′ nucleotides of the mature miRNA sequence of interest to complete the stem–loop oligonucleotide. The following is the stem–loop RT oligonucleotide for the mouse miR-495-3p: 5′-GTC GTA TCC AGT GCA GGG TCC GAGGTA TTC GCA CTG GAT ACG AC <u>**AAG AAG**</u>-3′ [5]. The full list of sequences is shown in Table 1.
2. Design a forward oligonucleotide, which is specific for the miRNA of interest and complementary to the 3′ end of the miRNA.

3. Design a reverse oligonucleotide by using the same 44 nucleotides of the stem–loop sequence for all the RT oligonucleotides. The following is the reverse oligonucleotide: 5′-GTG CAG GGT CCG AGG T-3′.

3.4 Stem–Loop RT Oligonucleotides Folding

1. Order HPLC-purified DNA oligonucleotides and verify concentrations. Refold stem–loop oligonucleotides as follows: add 30 μL of 10 μM stock in a new tube and heat to 95°C for 10 min. Reduce heat slowly to 75°C. Hold temperature at 75°C, 68°C, 65°C and 62°C for 1 h each, then hold temperature at 60°C for 2 h more.
2. Store structured oligonucleotides at −20°C.

3.5 microRNA Reverse Transcription

1. Per each miRNA of interest, prepare a 20 μL RT reaction as follows: 1 μL of 10 μM folded stem–loop oligonucleotide, 1 μL of 10 μM dNTPs mix, 4 μL buffer first strand 5×, 2 μL of 0.1M DTT, 1 μL M-MLV RT (200 U/μL), and 125 ng of miRNA. Bring up to 20 μL with ultra-pure water.
2. Place the tubes in a thermal cycler and set the reaction with the following program: 30 min at 16°C, 30 min at 42°C, and 5 min at 85°C.

3.6 End-Point PCR Analysis

1. For each of the miRNAs to be analyzed, prepare a 10 μL PCR mix as follows: 1 μL of buffer PCR 10×, 0.3 μL of 50 mM MgCl2, 0.2 μL of 10 mM dNTPs, 0.25 μL of 10 mM forward primer, 0.25 μL of 10 mM reverse primer, 0.2 μL of Taq polymerase and 250 ng of cDNA.
2. Place the reactions in a preheated thermal cycler and incubate 5 min at 94°C followed by 35 cycles of 94°C for 45 s; 62°C for 30 s and 72°C for 45 s. Include a final incubation of 7 min at 72°C.
3. Analyze reaction products by electrophoresis on a 1.5% agarose gel (Fig. 1b).

3.7 Expression Analysis of miRNAs by Quantitative Real Time PCR (qPCR) Assays

1. Aliquot 2× KAPA SYBR® FAST qPCR Master Mix according to manufacturer's instructions.
2. Prepare a PCR master mix by scaling the volumes according to the desired number of amplification reactions, including technical replicates and nontemplate negative controls (Fig. 1c, d) (*see* **Notes 5** and **6**).
3. For each of the miRNA to be analyzed, prepare a PCR master mix as follows: 10 μL of 2× KAPA SYBR® FAST qPCR Master Mix, 0.4 μL of 10 μM forward oligonucleotide, 0.4 μL of 10 μM reverse oligonucleotide, 1 μL of cDNA (250 ng). Bring the reaction up to 20 μL with ultra-pure water.

4. Run the amplification program as follows: 1 cycle of 10 min at 95°C, 40 cycles of 15 s at 95°C, and 1 min at 60°C.
5. For melting curve analysis, denature samples at 95°C, then cool to 65°C at 20°C per second. Collect fluorescence signals at 530 nm wavelength continuously from 65°C to 95°C at 0.2°C per second (*see* **Notes** 7 and **8**).
6. Analyze results using Q-Rex software (*see* **Note 9**).

4 Notes

1. All oligonucleotides are ordered desalted and HPLC purified.
2. It is very important that all samples are kept in ice during the rest of the procedure.
3. When a small amount of the tissue is available, an extra elution of the sample can be done, repeating the addition of 40 μL of Ultra-pure water in the center of the column, followed by a 2-min incubation at room temperature and centrifugation at 12,000 rpm for 30 s at 4°C.
4. Ensure ethanol is added to the RPE Solution before use.
5. To avoid cross-contamination within the samples and the control, the use of filter tips is mandatory.
6. Store kit protected from light at −20°C in a constant-temperature freezer.
7. To identify and validate the specificity or the amplification of the target. It is important that all amplicon products show a single curve and the same melting temperature.
8. For correct results, the threshold line must be set at the linear amplification area.
9. Apply the $2{-}\Delta\Delta CT$ to analyze the data. This method allows the calculation of relative gene expression levels through the threshold cycles (CTs) generated by the qPCR system, this analysis normalize first to an endogenous target (ΔCT), and then normalizes to a reference sample (ΔΔCT). The fold change between target samples is calculated with $2^{(-\text{delta delta CT})}$ [6] (fig. 5).

Conflict of Interest

The authors have no conflicts of interest to declare that are relevant to the content of this chapter.

References

1. Aranega AE, Lozano-Velasco E, Rodriguez-Outeiriño L, Ramirez de Acuña F, Franco D, Hernandez-Torres F (2021) MiRNAs and muscle regeneration: therapeutic targets in Duchenne muscular dystrophy. Int J Mol Sci 22(8):4236
2. Mousa NO, Abdellatif A, Fahmy N, El-Fawal H, Osman A (2023) MicroRNAs as a tool for differential diagnosis of neuromuscular disorders. Neuromol Med 25(4):603–615
3. Aksu-Menges E, Akkaya-Ulum YZ, Dayangac-Erden D, Balci-Peynircioglu B, Yuzbasioglu A, Topaloglu H, Talim B, Balci-Hayta B (2020) The common miRNA signatures associated with mitochondrial dysfunction in different muscular dystrophies. Am J Pathol 190(10):2136–2145
4. O'Brien J, Hayder H, Zayed Y, Peng C (2018) Overview of MicroRNA biogenesis, mechanisms of actions, and circulation. Front Endocrinol (Lausanne) 9:402
5. Xie S, Zhu Q, Qu W, Xu Z, Liu X, Li X, Li S, Ma W, Miao Y, Zhang L, Du X, Dong W, Li H, Zhao C, Wang Y, Fang Y, Zhao S (2019) sRNAPrimerDB: comprehensive primer design and search web service for small non-coding RNAs. Bioinformatics 35(9):1566–1572
6. Livak KJ, Schmittgen TD (2001) Analysis of relative gene expression data using real-time quantitative PCR and the 2(-Delta Delta C(T)) Method. Methods 25(4):402–408

Chapter 14

Gene Expression Analysis on Single Myofibers from Mouse Edl and Soleus Muscles

Amairani Cancino-Bello, Mauricio Hernández-Somilleda, Estela G. García-González and J. Manuel Hernández-Hernández

Abstract

Skeletal muscle is composed of different myofiber types with specific transcriptomic, proteomic, and metabolic identities. As such, skeletal muscle represents a valuable model to study cell differentiation and gene expression regulation. However, many molecular analyses of skeletal muscle are performed on whole muscle, with the caveat that many different cell types that constitute this organ contribute to the experimental results. Here, we describe a method to efficiently isolate individual myofibers from two muscle types and provide a guide to perform quantitative gene expression analysis of representative myosin heavy chain genes.

Keywords Skeletal muscle, Myofibers isolation, Soleus, Extensor digitorum longus, Eene expression, *Myh4*, *Myh7*, qPCR

1 Introduction

Skeletal muscle is composed of different cell types, including connective tissue, blood, nerves, vessels, and myofibers [1]. This cellular composition contributes to experimental results and interpretation of data when used to address molecular analyses [2, 3]. Although numerous studies have been published providing transcriptomic differences between oxidative and glycolytic muscles, gene expression patterns of isolated myofibers remain largely unidentified in many of these reports [4]. This protocol outlines a technique for isolating individual myofibers from the glycolytic, extensor digitorum longus (EDL) muscle, and the oxidative Soleus muscle. This method enables efficient RNA purification suitable for diverse downstream applications such as quantitative gene expression analysis, RNA-seq, ATAC-seq, and proteomics.

Paul Delgado-Olguín (ed.), *Mouse Embryogenesis*, Methods in Molecular Biology, vol. 3010,
https://doi.org/10.1007/978-1-0716-5126-1_14,

2 Materials

All the solutions listed in this section must be prepared before the dissection.

2.1 Dissection Materials and Reagents

1. Sterile dissecting scissors.
2. Sterile tweezers.
3. Ultrafine syringe 6 mm.
4. Autoclave.
5. 70% ethanol.
6. Sterile phosphate-buffered saline (PBS), pH 7.4, 1×.

2.2 Myofibers Isolation Materials and Reagents

1. Stereo microscope.
2. Sterile 100 × 25 × 25 mm tissue culture dish.
3. Glass Pasteur pipets.
4. Glass cutting pencil.
5. 37°C water bath.
6. Sterile horse serum (HS).
7. Humidified 37°C, 5% CO_2 incubator.
8. Collagenase solution: 2 mg/mL of Collagenase type I in DMEM (Dulbecco's modified Eagle's medium; 1 g/L glucose, L-glutamine with 110 mg/L sodium pyruvate), filtered through a 0.22 μm filter.
9. Washing medium: DMEM filtered through a 0.22 μm filter.

2.3 RNA Extraction and Quantification Materials and Reagents

Stock solutions should be diluted to the final concentration with ultrapure-free nuclease water.

1. TRIzol®.
2. Sterile dissecting scissors.
3. RNase-free disposable pellet pestles.
4. Chloroform, molecular biology grade.
5. RNase-free 1.5 mL centrifuge tubes.
6. High-speed refrigerated centrifuge.
7. Ethanol, molecular biology grade.
8. EZ-10 spin column total RNA Miniprep Kit.

2.4 cDNA Synthesis Materials

1. Oligo $(dT)_{20}$ (500 μg/mL). Store at −20°C.
2. Previously purified RNA from the myofibers of each muscle.

3. 2.5 mM dNTPs Mix, adjusted to a final concentration of 2.5 mM each dATP, dGTP, dCTP, dTTP.
4. Buffer first strand (5×).
5. DTT 100 mM.
6. M-MLV RT (200 U/μL).
7. Thermocycler for PCR.
8. PCR Master Mix (2×).

2.5 qPCR Quantification Materials

1. SYBR Green Master Mix (2×).
2. Amplification primer Forward (10 μM).
3. Amplification primer Reverse (10 μM).
4. cDNA previously synthesized from the myofibers of each muscle.
5. Thermocycler for real-time PCR.

3 Methods

The muscle dissection and myofiber isolation were performed as previously reported with certain modifications [5, 6].

3.1 Before Skeletal Muscle Dissection

Perform the following steps for EDL and Soleus separately.

1. Before dissection of EDL and Soleus, sterilize all the tools (scissors, tweezers, etc.) (*see* **Note 1**).
2. Coat 1 sterile 100 mm tissue culture dish with enough sterile HS serum until it covers the entire surface. Incubate at room temperature for 15 min, and then remove the HS and allow it to air dry for approximately 45 min. When dishes are completely dry, add 10 mL of washing medium and keep them in the incubator at 37°C and 5% CO_2 (*see* **Note 2**).
3. For myofibers isolation, cut one Pasteur pipette with a glass cutting pencil to obtain a size of 4–5 mm diameter for each. Coat each pipette with HS prior to use (*see* **Note 3**).
4. For myofibers manipulation, flame the opening of a normal Pasteur pipette and flame the middle to curve it. This will make it easier to transfer the myofibers.
5. Prewarm the collagenase solution at 37°C.

3.2 EDL and Soleus Dissections

1. Perform mouse euthanasia following institutional guidelines. For this experiment, we used an 8–12 weeks old mouse C57BL6J.

2. Use 70% ethanol to disinfect the limbs of the mouse. Place the mouse face up on a dedicated support.
3. Use scissors to cut the skin around the ankle and up to the knee. With the tweezers, remove the skin to expose the muscles.
4. Use scissors and tweezers to remove the fascia that surrounds the tibialis anterior (TA), cutting along the tibia bone from the ankle up to the knee. Do this carefully (*see* **Note 4**).
5. Locate the distal tendons of the extensor digitorum longus (EDL) and TA muscles; they are located very close to each other. For TA removal, insert an ultrafine syringe between the two distal tendons (under the TA tendon) and separate both muscles, sliding the needle towards the knee. Cut the distal TA tendon and then cut the proximal TA tendon as near to the patella as possible. At this point, the EDL is entirely visible.
6. The next step is to identify the proximal EDL tendon; for this, you must cut the biceps femoris posterior (BFP) as close as possible to the patella. The proximal EDL tendon must be visible at this point (*see* **Note 5**).
7. Insert the ultrafine syringe under the proximal EDL tendon and cut it with sterile scissors. Slide the ultrafine syringe to the ankle. This will allow you to obtain an entire EDL. Cut the distal EDL tendon to liberate it from the bone (*see* **Note 6**).
8. Repeat steps 3 through 10 to isolate the EDL muscle in the hind limb on the opposite side (Fig. 1a).
9. For Soleus dissection, place the mouse face down on the support. Be sure to remove all surrounding fascia as well. Locate the distal tendons of both gastrocnemius (GAS) and Soleus (SOL) muscles and insert an ultrafine syringe under the tendons. Slide the syringe gently as close as possible to the distal tendons to separate the tissues from the bone.
10. Insert the syringe between the proximal GAS tendon and proximal SOL tendon and slide gently to the distal tendon to separate both muscles.
11. Cut both the proximal tendons to see the SOL muscle. You can distinguish SOL by its intense red color in comparison with the pink color of the gastrocnemius.
12. Take the proximal Soleus tendon with tweezers, lift it, and cut the distal tendon as close as possible to the knee to obtain the whole muscle from tendon to tendon (*see* **Note 7**).
13. Repeat steps 9 through 12 to isolate the SOL muscle in the hind limb on the opposite side (Fig. 1c).

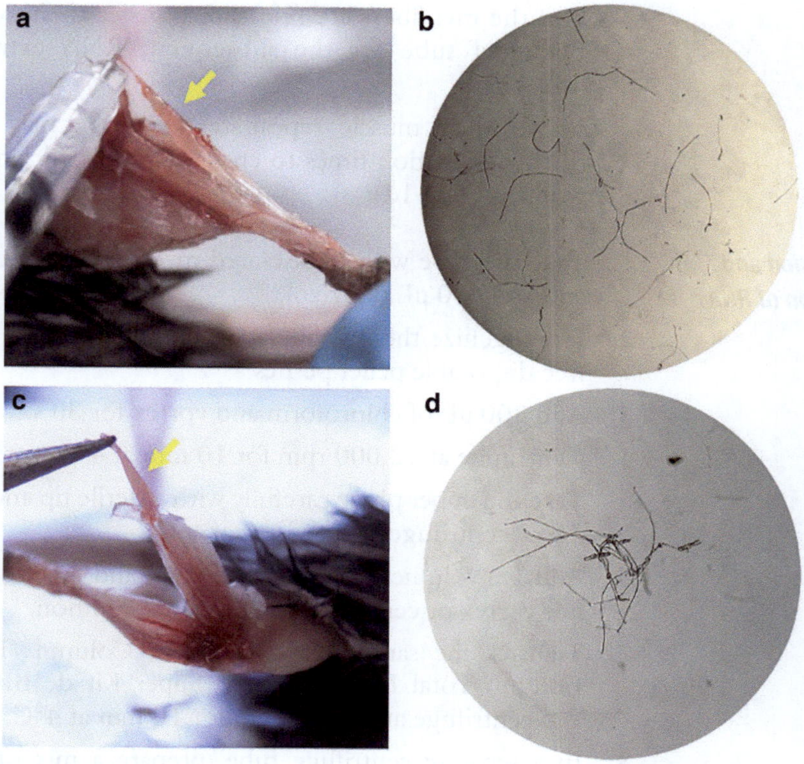

Fig. 1 Dissection of the extensor digitorum longus (EDL) and SOL muscles and myofiber isolation. At this stage, the skin and fascia were removed. (**a**) The EDL muscle before the distal tendon cut, marked by the yellow arrow. (**b**) Isolation of EDL myofibers observed under a stereo microscope. (**c**) SOL muscle dissection before proximal tendon cut, indicated by the yellow arrow. (**d**) SOL myofibers isolation observed under a stereo microscope. Images (**b**) and (**d**) were captured at the same magnification, but (**d**) had to be enlarged for better visibility

3.3 Digestion and Myofibers Isolation

Proceed with the following steps for EDL and Soleus separately.

1. Transfer both extensor digitorum longus (EDL) muscles into a 15 mL Falcon tube containing 2 mL of Collagenase solution and incubate in a 37°C water bath for 1 h. Gently mix the tube every 15 min. You will notice that the muscles begin to increase in size and will have the appearance of hair, which will be the myofibers (*see* **Note 8**).
2. Stop digestion by transferring the digested muscles to an HS-coated 100 mm tissue culture dish containing 10 mL of washing medium. The subsequent procedures are conducted utilizing a stereo dissecting microscope with transmitted bright field illumination (Fig. 1b).
3. After transferring the muscles onto the dish, break them up using a 4–5 mm diameter pipette. Employ gentle pipetting motions, moving up and down against the edge of the dish, to facilitate the release of myofibers (*see* **Note 9**).

4. Once the myofibers are released, transfer them to a 1.5 mL Eppendorf tube for the subsequent RNA extraction (*see* **Note 10**).
5. For the Soleus muscle, repeat steps 1–4. You may have to adjust the incubation times to ensure a more efficient release of myofibers (Fig. 1d).

3.4 Extraction and Quantification of RNA

1. Take the tube with the isolated myofibers from each muscle and add 600 μL of Trizol.
2. Homogenize the myofibers inside the tube with the RNase-free disposable pellet pestles.
3. Add 200 μL of chloroform and vortex for 30 s.
4. Centrifuge at 12,000 rpm for 10 min at 4°C (*see* **Note 11**).
5. Take the upper phase carefully with a sterile tip and transfer to a new centrifuge tube.
6. Add 1.5 volumes of absolute ethanol and pipette to mix. Do not vortex or centrifuge after ethanol addition.
7. Transfer the sample to a TR Spin Column (EZ-10 Spin Column Total RNA Miniprep Super Kit de BIOBASIC™) and centrifuge at 12,000 rpm for 10 min at 4°C.
8. In a separate centrifuge tube, prepare a mix of 5 units of DNase with 25 μL of the RPE solution (per sample).
9. Carefully transfer the DNase mix directly to the column. Incubate at room temperature for 15 min.
10. Add 500 μL of RPE solution to the column and centrifuge at 12,000 rpm for 1 min.
11. Discard the flow through and repeat the previous step.
12. Centrifuge again at 12,000 rpm for 1 min to completely dry the column (*see* **Note 12**).
13. Change the column to a new centrifuge tube, add directly 40 μL of ultrapure-free nuclease water, and incubate at room temperature for 2 min.
14. Centrifuge at 12,000 rpm for 1 min to elute the RNA.
15. The purified RNA should be quantified and stored at –80°C (*see* **Note 13**).

3.5 Synthesis of cDNA

The next volumes are adjusted for a single reaction of 20 μL for the synthesis of single-strand cDNA (*see* **Note 14**).

1. Prepare 2 mixes, A and B, in separate microcentrifuge tubes.

2. For Mix A, use the 250 ng aliquot of RNA and mix 1 μL of Oligo $(dT)_{20}$ (500 μg/mL) and 1 μL of 2.5 mM dNTPs Mix; the final volume should be 12 μL.
3. Add 4 μL of buffer first strand 5×, 2 μL DTT 100 mM, and 1 μL of free nuclease water to Mix B.
4. Mix A should be heated to 65°C for 5 min. Immediately after incubation, chill on ice.
5. Add the contents of Mix B to the tube of Mix A. Incubate for 2 min on ice.
6. Add 200U of M-MLV RT to the mixture and pipette gently. Incubate at 37°C for 50 min.
7. After incubation, heat to 70°C for 15 min to inactivate the enzyme.
8. Dilute the reaction product with ultrapure-free nuclease water up to 40 μL final volume. The final concentration of cDNA should be 6.25 ng/μL.
9. The single-strand cDNA is ready to use for PCR or qPCR. Store at −20°C (*see* **Note 15**).

3.6 Quantitative PCR Analysis of Specific Myofibers Genes

Before starting (*see* **Note 16**).

1. Prepare a mixture for each reaction with a technical replicate. It is necessary to run a qPCR reaction of each sample with a housekeeping reference gene. Additionally, a no template control (NTC) reaction is needed for each gene as a negative control.
2. Set a 20 μL reaction containing the following components: 10 μL of SYBR Green Master Mix (2×) with 0.4 μL of 10 μM Forward primer and 0.4 μL of 10 μM Reverse primer (*see* **Note 17** and Table 1), add 7.2 μL of Ultrapure free nuclease water. At the end, add 2 μL of the cDNA previously synthesized and pipette up and down gently (*see* **Note 18**).
3. Cap the reaction plate/tube.
4. Select the correct qPCR protocol for the equipment and run (*see* **Note 19**).
5. Analyze using the appropriate guidelines for the instrument used. In this case, the data was analyzed and compared using the Relative gene expression method reported previously by Livak et al. [7] (Figs. 2 and 3).

Table 1
Real-Time PCR primers

Primers	Sequences 5′–3′	Gene target	Species
Myh7 Fwd Myh7 Rev	ACCTACTCGGGCCTCTTCTGC AGGGGCCTCGCTCCTCTTCTT	Myh7; Myosin Heavy Chain 7	Mouse
Myh4 Fwd Myh4 Rev	TCAATGAGATGGAGATCCAGCTGAAC GTCCAGGTGCAGCTGTGTGTCCTTC	Myh4; Myosin Heavy Chain 4	Mouse
Tbp Fwd Tbp Rev	CAAGTTTACACGCAAGATTCACG CACCAATGACTCCTATGACCC	Tbp; TATA-Box Binding Protein	Mouse

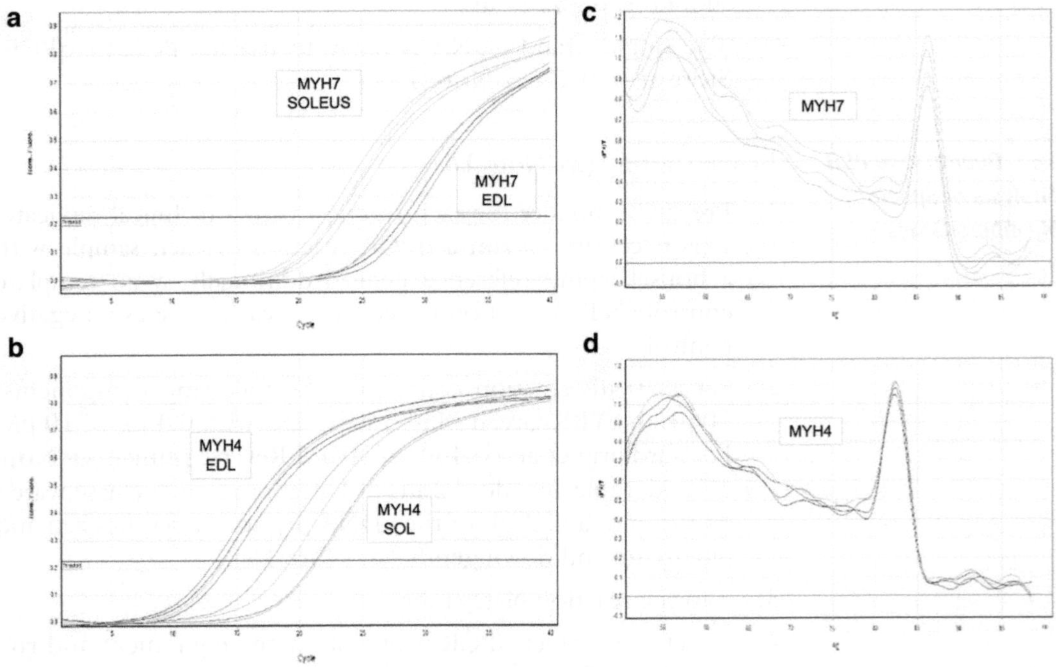

Fig. 2 Dissociation curves and melting curves of real-time SYBR green PCR analysis for the extensor digitorum longus (EDL) and SOL muscles. **a**) Melting curve for *Myh7*. *Myh7* is a gene more prominently expressed in the SOL muscle, and fluorescence detection commences earlier in this sample, as expected. (**b**) Melting curve for *Myh4*. *Myh4* is a gene more prominently expressed in the EDL muscle, and fluorescence detection begins earlier compared to SOL, consistent with expectations. In images (**c**) and (**d**), a single melt peak indicates the amplification of a single PCR product in the samples

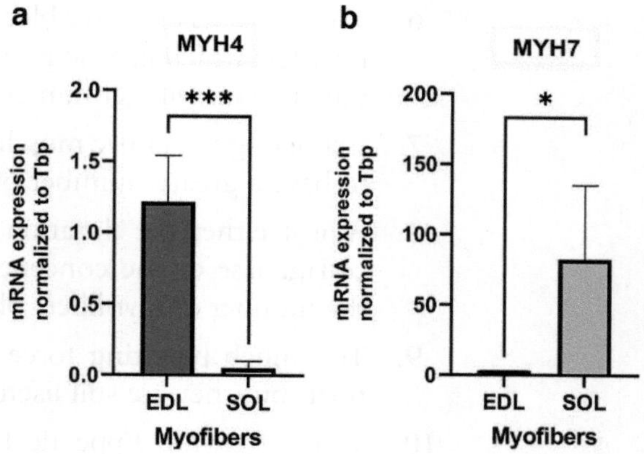

Fig. 3 Relative expression of mRNA for myosin heavy chain isoform (*Myh4* and *Myh7*) in the extensor digitorum longus (EDL) and Soleus muscles. Values are calculated based on the $2^{-\Delta\Delta CT}$ method (where Ct is the threshold cycle), which gives the relative target gene expression compared to Tbp expression. (**a**) Differences in relative *Myh4* mRNA expression levels between the extensor digitorum longus (EDL) and Soleus (SOL) muscles are statistically significant. The levels of *Myh4* in the extensor digitorum longus (EDL) muscle are greater than those observed in the Soleus (SOL) muscle. (**b**) Differences in relative *Myh7* mRNA expression levels between the EDL and SOL are statistically significant. The levels of *Myh7* in the SOL are greater than those observed in EDL, as we expected. $n = 4$ biological replicates. Error bars = standard error of the mean. Different superscripts indicate a significant difference (*$p < 0.05$, ***$p < 0.001$, *t*-student two-tailed test)

4 Notes

1. Make sure the tools for dissection have been sterilized in the autoclave at 121°C before you start the dissection.
2. It is crucial to coat the tissue culture dishes and Pasteur pipettes with HS to prevent myofibers from sticking to the dish and/or pipette.
3. Flame and polish the pipette correctly to avoid damaging the muscle and myofibers. Must cut and prepare a minimum of 4–5 pipettes in case they break during the experiment.
4. Remove the muscle fascia to avoid muscle damage. If you skip this step, the myofibers will contract.
5. Cutting the biceps femoris posterior (BFP) correctly is necessary to make visible the proximal EDL tendon. Cutting from the distal tendon to the proximal tendon (in both cases, EDL and SOL) is a critical step for obtaining noncontracted myofibers. You may choose to cut either the proximal tendon or the distal tendon first; there is no strict guideline for this decision.

6. This prevents excessive bleeding during the dissection of both muscles, which can be a problem because, by damaging the muscle, you will get numerous contracted myofibers.
7. Extracting the entire muscle from tendon to tendon is crucial to have a greater number of myofibers.
8. Adjust either the duration the Soleus muscle is exposed to collagenase or the concentration of collagenase to optimize the number of myofibers obtained.
9. Too much pipetting force will cause the myofibers to contract, but they are still useful for RNA extraction.
10. Transfer to the Eppendorf tube both living and contracted myofibers, contracted myofibers usually have some nuclei. Keep the tubes on ice. Remember to coat every glass pipette or plastic pipette tip with HS serum to prevent the sticking of myofibers.
11. After centrifugation, the samples will be separated into 3 phases. The colorless upper phase contains the RNA. Do not take another phase to avoid protein or DNA contamination.
12. This step is very important to remove all the residual ethanol.
13. After RNA purification, it should be handled on ice to avoid its denaturation. Once it is quantified, it should be distributed to nuclease-free microcentrifuge tubes on 250 ng RNA aliquots, adjusting to a final volume of 10 μL with ultrapure water. These tubes are ready to use for cDNA synthesis to avoid RNA denaturation after multiple thaws from the stock solution. It is imperative to verify RNA integrity through agarose gel electrophoresis before proceeding with cDNA synthesis.
14. Solutions should be completely thawed for use. While using, keep them on ice. Refreeze immediately after use at −20°C.
15. The synthesized cDNA should be tried before use on qPCR. The simplest way to prove that cDNA was generated is by running a PCR. This could be useful to prove that the primers are specific for the gene of interest and that the cDNA was correctly synthesized. PCR can be prepared according to Lee et al. [8].
16. Depending on the reference dye used, the manufacturer's instructions may change. In our instance, we used SYBR Green I dye with a Rotor-Gene Q thermocycler. For other equipment, it could be necessary the use of ROX dye.
17. Primers should be designed with a melting temperature of approximately 60°C. For optimal results, the length of the PCR product should be between 60 and 200 bp. The final concentration of the primers should be 200 nM and adjusted to this if the volume is modified.
18. Mix well every reaction component by pipetting. The amount of final cDNA added to the reaction must not exceed 20 ng

for a 20 μL. If the final reaction volume is modified, adjust proportionally.

19. The setup for the qPCR reaction was used according to the Rotor-Gene® Q real-time PCR manual. When using another thermocycler, adjust the setup according to the equipment.

Acknowledgments

We gratefully acknowledge the financial support provided by Consejo Nacional de Humanidades Ciencia y Tecnología through the grant FORDECyT-PTONACES/140637/2020.

Conflict of Interest The authors have no conflicts of interest to declare that are relevant to the content of this chapter.

References

1. Rubenstein AB, Smith GR, Raue U, Begue G, Minchev K, Ruf-Zamojski F, Nair VD, Wang X, Zhou L, Zaslavsky E, Trappe TA, Trappe S, Sealfon SC (2020) Single-cell transcriptional profiles in human skeletal muscle. Sci Rep 10(1):229
2. Lovrić A, Rassolie A, Alam S, Mandić M, Saini A, Altun M, Fernandez-Gonzalo R, Gustafsson T, Rullman E (2022) Single-cell sequencing deconvolutes cellular responses to exercise in human skeletal muscle. Commun Biol 5(1):1121
3. Cai C, Yue Y, Yue B (2023) Single-cell RNA sequencing in skeletal muscle developmental biology. Biomed Pharmacother 162:114631
4. Dos Santos M, Backer S, Saintpierre B, Izac B, Andrieu M, Letourneur F, Relaix F, Sotiropoulos A, Maire P (2020) Single-nucleus RNA-seq and FISH identify coordinated transcriptional activity in mammalian myofibers. Nat Commun 11(1):5102
5. Brun CE, Wang YX, Rudnicki MA, Single EDL (2018) Single EDL myofiber isolation for analyses of quiescent and activated muscle stem cells. Cells. Methods Mol Biol 1686:149–159
6. Blackburn DM, Lazure F, Corchado AH, Perkins TJ, Najafabadi HS, Soleimani VD (2019) High-resolution genome-wide expression analysis of single myofibers using SMART-Seq. J Biol Chem 294(52):20097–20108
7. Livak KJ, Schmittgen TD (2001) Analysis of relative gene expression data using real-time quantitative PCR and the 2(-Delta Delta C(T)) method. Methods 25(4):402–408
8. Lee PY, Costumbrado J, Hsu CY, Kim YH (2012) Agarose gel electrophoresis for the separation of DNA fragments. J Vis Exp 62:3923

Chapter 15

Characterization of Adipocyte Precursor Cells (APCs) by Lineage Tracing, Flow Cytometry, and Cytometry Time of Flight (CyTOF)

Qiwei Zhang, Nathaniel Vo, Ju Hee Lee, Azadeh Yeganeh, Kafi Ealey and Hoon-Ki Sung

Abstract

Flow cytometry and fluorescence-activated cell sorting (FACS) techniques have significantly advanced the characterization of adipocyte precursor cell (APC) populations. They allow immuno-phenotyping, quantification, and isolation of distinct populations, which critical for understanding adipose tissue development and homeostasis. Here, we describe the identification and purification of APCs using flow cytometry, FACS, and cytometry by time-of-flight (CyTOF), defined by previously established surface marker profiles. In addition, we describe the mouse models and whole adipose tissue visualization techniques that will enable us to characterize the cellular origin and plasticity of APCs.

Keywords Flow cytometry, CyTOF, Adipocyte precursor cells (APCs), Lineage tracing, Cre-recombinase transgenic mouse model, Adipose biology

1 Introduction

Adipose tissue is a loose connective tissue found ubiquitously throughout the body. It is traditionally classified into two types: white adipose tissue (WAT), which stores excess energy in the form of triglycerides, and brown adipose tissue (BAT), which generates nonshivering thermogenesis [1]. The current epidemic of obesity and associated metabolic diseases, such as type 2 diabetes, demands a better understanding of how adipocytes develop and where they originate, to answer fundamental questions about adipose tissue biology.

Adipose tissue depots are primarily distinguished by their anatomical locations. Two major types of WAT are dispersed throughout the body: visceral WAT, distributed around internal organs,

Paul Delgado-Olguín (ed.), *Mouse Embryogenesis*, Methods in Molecular Biology, vol. 3010,
https://doi.org/10.1007/978-1-0716-5126-1_15,

and subcutaneous WAT, found underneath the skin [2]. Although WAT begins to develop in utero, the majority of its development and expansion occurs after birth and continues throughout the lifetime [3]. Unlike WAT development, which occurs predominantly in the postnatal period, BAT development occurs before birth [4]. In humans, a large depot of BAT is present in newborns, but only small amounts persist throughout adulthood. In rodents, BAT is mostly distributed in the interscapular and perirenal regions [1, 5]. Brown adipocytes express high levels of the mitochondrial protein uncoupling protein-1 (UCP1), due to their requirement for high amounts of ATP to carry out nonshivering thermogenic processes [6]. Beige/brite adipocytes are brown-like adipocytes that show distinct characteristics from WAT and BAT. They reside within WAT, and are activated upon various stimuli, such as cold exposure, β3-adrenergic activators, exercise, and fasting [7–9]. WAT and BAT develop from the mesoderm, but they are derived from distinct precursor populations. Mesenchymal stem cells committed to the Myf5-negative lineage primarily give rise to white preadipocytes whereas Myf5-positive stem cells give rise to BAT [10]. Several adipogenic transcriptional cascades, such as peroxisome proliferator-activated receptor-gamma (PPARγ) and CCAAT/enhancer-binding protein alpha (C/EBPα), regulate differentiation of WAT and BAT [10]. Furthermore, in response to thermogenic stimuli such as cold exposure or β3-adrenergic stimulation, white adipocytes can convert to beige adipocytes through the process of transdifferentiation [10]. The bi-directional interconversion of white and beige adipocytes, and the adaptive physiological response of WAT under environmental challenges, sparks an interest in the development and metabolic function of beige adipocytes and their therapeutic potential for obesity and diabetes.

Here, we describe experimental methods for the identification and purification of adipocyte precursor cells (APCs) using flow cytometry, FACS, cytometry by time-of-flight (CyTOF) [11, 12], and whole adipose tissue visualization. These techniques differ in their application to adipose tissue analysis: flow cytometry provides basic cell population sorting and marker detection [13], CyTOF enables high-dimensional single-cell analysis with multiple markers simultaneously [11], and FACS combines cell sorting with fluorescence detection for isolating specific adipose cell populations based on their phenotype [14]. The combination of these methods enables precise adipose tissue cell sorting, high-dimensional profiling, and detailed functional analysis at single-cell resolution. In addition, we describe the mouse models that will enable us to characterize the cellular origin and plasticity of APCs.

2 Materials

Prepare and store all reagents at 4°C unless indicated otherwise.

2.1 Single Cell Preparation of SVF Cells from Mouse Adipose Tissue

1. Tissue digestion cocktail: Dissolve 1 mg/mL collagenase II in DMEM/F12 (1:1) media and mix well. Add 5 mL fatty acid-free BSA and 1 mL of 500 μg DNase I. Filter the cocktail through a 0.2 μm filter. Make aliquots and store at −80°C until use.
2. Tissue storage solution (Miltenyi) (optional).
3. Rotating cell shaker/incubator at 37°C (or 37°C water bath).
4. Wash buffer: DMEM/F12 media, 10% FBS.
5. Sterile petri dish and surgical tools.
6. Cell strainers: 40, 70, and 100 μm.
7. Red blood cell lysis solution.

2.2 Cell Staining and Analysis for Flow Cytometry

1. Flow staining buffer: 1× HBSS with Ca^{2+}/Mg^{2+}, 10 mM NaN_3, 10 mM HEPES solution, 2% FBS.
2. FcR blocking solution: CD16/CD32 antibody (1:200 in flow staining buffer).
3. Fixable viability dye.
4. Antibodies of target markers tagged with fluorophore-conjugated secondary antibodies, prepared in flow staining buffer. 2× Antibody master mix should be prepared in advance.
5. 70 μm-filtered FACS tubes.
6. Compensation beads.
7. Fixation buffer for optional fixation and storage of cells prior to flow cytometry analysis: 2% PFA/PBS.

2.3 Cell Sorting and Adipocyte Differentiation

1. Cell collecting media: DMEM/F12 media, 50% FBS. Aliquot into sterile Eppendorf tubes and store until use.
2. Culture plating media: DMEM/F12 media, 10% FBS, 1% penicillin/streptomycin.
3. Adipocyte differentiation media: DMEM/F12 media, 10% FBS, 0.25 μM dexamethasone, 1 μM insulin, 0.5 mM isomethylbutylxanthine.
4. Multi-well cell culture plates.

2.4 Cell Preparation for CyTOF Analysis

1. CyTOF staining media (CSM): PBS + 1% BSA (~15 mL/sample).
2. FcR blocking antibody mix: 1:200 in CSM.

3. Prepare in advance: Antibody cocktail: 2× surface antibody cocktail master mix (for 1–3 million cells), prepared in CSM. CyTOF antibody panels can contain more than 40 markers, and it can be time-consuming to prepare a master antibody mix on the day of the sample preparation. We routinely prepare the antibody mix 24–48 h prior to antibody staining and store at 4°C. All antibodies should be previously titrated to determine the optimal dilutions for staining.
4. Cisplatin viability stain.
5. 2× Cisplatin solution: 10 mL PBS + 2 μL cisplatin stock.
6. Methanol-free 16% ultrapure formaldehyde (FA).
7. Fixation media: PBS + 0.3% Saponin + 1.6% FA, filtered.
8. Iridium Cell-ID™ Intercalator (Standard BioTools –500 μM stock).
9. Iridium staining media: 10 mL Fixation media + 10 μL iridium intercalator.
10. FBS.
11. DMSO for optional long-term storage of stained samples prior to acquisition.
12. 4 mL conical tubes (Diamed STK 8550).
13. Cell-ID™ 20-plex Pd barcoding kit (Standard BioTools) (optional).

2.5 Preparation for Whole Mount Staining of Adipose Tissue and Microscopy

1. Fixation media for whole-mount imaging: Prepare a fresh batch of 1% paraformaldehyde (PFA) in 1× PBS before each use.
2. Blocking media: Add 500 μL of goat serum plus 10 μL of Triton X-100 in 10 mL of 1× PBS and mix well.
3. Mounting media with DAPI.

3 Methods

3.1 Using Cre/lox System to Study Adipose Biology

The Cre-loxP recombination system is a tool used for site-specific manipulation of genes of interest, where Cre expression can be targeted at a cell-/tissue-specific level. This system has revolutionized the study of the function and the development of adipose tissue in vivo. The development of in vivo Cre induction by administration of tamoxifen or doxycycline allows the observation of spatiotemporal regulation of a target gene [15]. Several adipocyte-specific promoters have been generated to express Cre recombinase only in adipose tissues. Some of the commonly used adipose-specific Cre transgenes include aP2 (Fabp4), AdipoQ, Ng2, Pparγ, PDGFR-α, and PDGFR-β [16–18]. Various fluorescent reporters are combined with Cre transgenes to define Cre activity in a spatiotemporal manner.

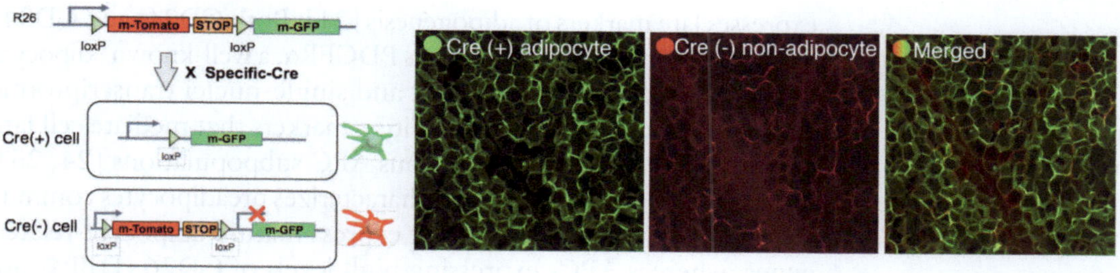

Fig. 1 Lineage tracing system for adipose tissue. Cre(+) cells express mGFP by excising two loxP sites, while Cre(−) cells express mTomato

3.2 Using In Vivo Lineage Tracing to Track Adipocyte Precursor Cells

Lineage tracing technology has long been used to identify the progeny of a single cell and has provided a better understanding of developmental dynamics and the fate and plasticity of stem cells [19]. In combination with cell-type-specific Cre-recombinase (Cre)-expressing models, and multicolor reporter mouse lines (e.g., Brainbow and mT/mG mouse), lineage tracing technology has provided unprecedented insight into various scientific questions, including the origin of an adipocyte and tracking of adipocyte precursor populations [16, 18, 20, 21]. In this system, adipose-specific expression of Cre turns on the fluorescent reporter protein in a Cre-expressing cell-specific manner, allowing visualization/tracing of the adipocyte progenitor and their descendant adipocytes.

For studies focused on mature white adipocytes, the traditional reporters that mainly stain the cytoplasm are difficult to visualize due to the lack of cytoplasm in mature adipocytes. To overcome this limitation, mT/mG (membrane-tdTomato/membrane-GFP), a reporter that marks Cre excision by switching from membrane-targeted tdTomato to membrane-targeted GFP, has become a commonly used genetic tool in adipose biology. It is a double-fluorescent system that allows direct visualization of Cre-excised and nonexcised cells at a single cell resolution (Fig. 1). The localization of fluorescent proteins to the membrane structures allows visualization of the entire adipocyte morphology, making it an ideal reporter system for studying adipocytes [21, 22].

3.3 Characterization of Adipose Precursor Populations Using Flow Cytometry, Fluorescence-Activated Cell Sorting (FACS), and Cytometry Time Of Flight (CyTOF) Analysis

APC populations from mice have been well characterized using flow cytometry and fluorescence-activated cell sorting (FACS) to isolate populations of interest based on expression of cell surface markers. Initially, using FACS, two precursor cell populations from WAT were identified in mice, characterized by negative expression for hematopoietic and endothelial-specific lineage markers (CD45, Ter119, CD31) and positive expression of mesenchymal and stem cell markers (CD29, CD34, Sca1) [23]. Further separation of these populations based on expression of CD24 revealed adipogenic commitment. Loss of CD24 expression characterized preadipocytes that are less proliferative and more committed to the adipogenic lineage. Transplantation studies showed that in vivo, CD24-expressing adipocyte progenitors generate the CD24^{-} preadipocyte population, which expresses late markers of adipogenesis [21]. Both $CD24^+$ and $CD24^-$ cell populations commonly express PDGFRα, a well-known adipocyte lineage marker [21]. Single-cell and single-nuclei transcriptomic studies have further identified unique markers that mediate cell fate decisions of distinct heterogeneous APC subpopulations [24, 25]. Expression of CD54 (ICAM-1) characterizes preadipocytes committed to the adipocyte lineage that express mature adipocyte-related genes, whereas APCs expressing high levels of CD26 (DPP4) are stem-like multipotent mesenchymal progenitor cells with high-proliferative capacity and poor adipogenic capacity [24]. In vivo, DPP4+ cells were shown to give rise to committed preadipocytes expressing ICAM-1 and CD142 [24]. Subsequent studies revealed that high expression of the tetraspanin CD9 identified preadipocytes with a profibrotic cell fate [26].

Several studies have demonstrated the importance of angiogenesis in WAT development and expansion. Adipogenesis is tightly associated with and modulated by angiogenesis in a spatiotemporal manner throughout prenatal and adult life. Multiple studies have demonstrated the importance of adipocyte development in coordination with the vasculature [27–30]. It is suggested that perivascular mural cells (or pericytes) are a source of mesenchymal stem cells, and that a certain adipocyte population may be derived from a mural cell origin. For example, a study using PPARγ lineage tracing demonstrated that the vascular walls of adipose tissue blood vessels represent a niche for white adipocyte precursor cells. In particular, PDGFR-β marks mural cells, and the adipose PDGFR-β^+ SV cells displayed higher adipogenic potential than PDGFR-β^- SV cells [27].

Mass cytometry (CyTOF, Cytometry by Time-of-Flight Mass Spectrometry) is a high-throughput, high-dimensional analytical technology for characterizing cellular phenotypes and functions. Unlike conventional fluorescence-based flow cytometry, CyTOF employs antibodies conjugated with metal isotopes and detects their abundance through a mass spectrometer, eliminating issues of spectral overlap [12]. In adipose tissue research, CyTOF serves as a powerful tool to elucidate the composition and function of immune cells, adipocytes, and other stromal cells in the tissue microenvironment [11]. By employing multiple metal isotope labels, CyTOF enables the classification and analysis of complex cellular populations using diverse antibodies. Thus, utilizing commonly used markers for APCs, such as PDGFRα (CD140a), Sca-1, CD34, and CD29, enables the precise identification of APCs and their subpopulations, as well as the analysis of their differentiation potential and dynamic changes. Finally, by detecting functional markers such as CD24 and CD9, CyTOF reveals APC interactions within diverse microenvironments, providing critical support for studies on adipogenesis and cellular alterations in metabolic diseases [11].

3.4 Protocol for Identifying Adipose Progenitor Cells Using Flow Cytometry Analysis, and Fluorescence-Activated Cell Sorting (FACS)

Preparation of Single-Cell Suspension

1. Prepare tissue digestion cocktail as previously indicated (Cocktail solution should aliquoted and stored at −80°C). Thaw at 37°C immediately prior to use.
2. Harvest mouse adipose tissue depots (inguinal, perigonadal, perirenal, and mesenteric) and store on ice or in tissue storage solution, for longer-term storage (*see* **Note 1**).
3. Mince the tissue to a fine consistency (1–2 mm) to form a paste using sterile scissors (*see* **Note 2**).
4. Add approximately 10 mL of the digestion cocktail into the petri dish with the finely minced tissue. Suspend the minced tissue in the enzyme cocktail and transfer to a 50 mL tube.
5. Incubate for ~30–45 min in a rotating shaker at 37°C. In the last 5 min of digestion, shake the tube vigorously (5–10 times) by hand to assist breaking down the undigested tissue. If there is no access to a rotating shaker, incubate cells in a 37°C water bath and shake vigorously by hand every 15 min to ensure uniform tissue digestion.
6. After digestion is complete, add double the volume of wash buffer to the tissue digest to neutralize the enzyme and stop digestion. Filter through a 100 μm cell strainer into a new sterile 50 mL centrifuge tube. Wash the filter with 5–10 mL of the wash buffer. Discard undigested tissue on the strainer (*see* **Note 3**).
7. Centrifuge at 400 *g* for 10 min and aspirate the supernatant (including the floating top fraction containing mature adipocytes) without disturbing the bottom-layer containing the stromal vascular fraction (SVF) pellet.
8. Resuspend the pellet in 1–2 mL of red blood cell lysis solution.
9. Incubate at room temperature for 2 min.
10. Add 5–10 mL of wash buffer, centrifuge at 300 *g* for 5 min and discard the supernatant. If the cells still appear red, repeat steps 8–10.
11. Resuspend the pellet in 1 mL of wash buffer and filter the suspension through a 70 μM filter into a 5 mL FACS tube. Count cells. For flow cytometry and FACS, proceed to

the next section "Preparation of Cells for Flow Cytometry Analysis". For CyTOF cell preparation and analysis, proceed to section 3.5.

Preparation of Cells for Flow Cytometry Analysis

1. Wash 1–3 million cells in flow staining buffer by centrifugation at 300 *g* for 5 min.
2. Aspirate the supernatant and resuspend cells in ~200 μL of flow staining buffer.
3. Perform viability dye staining using a fixable viability stain (concentration of 1:500–1:1000) according to the manufacturer's protocol, particularly if cells will be fixed and stored for analysis on a later date. If cells will be analyzed by flow cytometry on the same day, a nonfixable viability dye such as propidium iodide can be added to the cells just prior to acquisition on the flow cytometer (*see* **Note 4**).
4. Add 1 mL of flow staining buffer to cells, centrifuge at 300 *g* for 5 min and remove the supernatant.
5. Add 25 μL of Fc-block solution (CD16/CD32) to the pelleted cells and incubate on ice for 10–15 min. Before proceeding to the following step, prepare an antibody staining cocktail containing all fluorochromes of interest for all experimental samples. In addition to the experimental samples, four sets of control samples are required:
   (a) FMO (fluorescence-minus-one) samples, including every antibody except one fluorochrome. For example, the FITC FMO sample will contain every antibody except the FITC antibody.
   (b) Single antibody staining, prepared with suitable compensation beads containing each antibody.
   (c) Unstained sample: Prepare one tube containing unstained cells.
   (d) Viability control, a tube containing cells stained with the viability dye (*see* **Note 5**).

   And a common adipose precursor cell population panel: CD31−, CD45−, Pdgfrα+, CD29+, CD34+, Sca1+, CD24+ (for adipocyte progenitor population) or CD24− (for preadipocyte population).
   (a) When using a mouseline with a fluorescent reporter, such as mT/mG mice, avoid using fluorochromes that overlap with the fluorescent reporter of the mouse.
   (b) Antibodies to CD26, CD54, and CD9 can also be used to further identify distinct stem-like, adipogenic, and fibrogenic cell populations, respectively.
6. Add 25 μL of prepared 2× antibody cocktail to cells to make a final 1× cell staining mix in 50 μL volume and incubate for 30 min at 4°C, protected from light.

7. Wash the stained cells with 1 mL of flow staining buffer, centrifuge for 5 min at 400 *g*, then remove the supernatant.
8. If flow cytometry analysis is performed on the same day, resuspend the cells in a final volume of 300–500 μL of flow staining buffer, then pipette the cells through 70 μm-filtered FACS tubes. Place on ice, protected from light until loading into the flow cytometer. Proceed to section Flow Cytometry Analysis. If cells cannot be analyzed in the flow cytometer on the same day, they can be fixed in 2% PFA/PBS and stored at 4°C for up to 1 week prior to analysis (see step #9).
9. For optional cell fixation, add 500 μL of 2% PFA/PBS to pelleted cells from step #8 and incubate at room temperature for 10 min. Wash cells in 2 mL PBS, resuspend in 1 mL of flow staining medium and store at 4°C for up to 1 week. Prior to flow cytometry analysis, centrifuge cells for 5 min at 400 *g*, remove the supernatant, resuspend in 300–500 μL of flow staining medium and proceed to the next section.

Flow Cytometry Analysis

Using a flow cytometer with 4–5 lasers and at least 10–12 detectors enables the simultaneous detection of at least 10 different fluorochromes.

1. Acquire the unstained cells as a negative control to eliminate the background signal on all the channels.
2. Next, acquire the unstained cells stained only with the viability marker.
3. Then, run the compensation beads stained with a single antibody to find the spectral overlap among the fluorochromes. These differences can be removed by changing the voltages or by compensation during the analysis.
4. Acquire the experimental tube stained with all the antibodies, followed by the FMO control tubes. (In multicolor flow cytometry, the FMO controls will be used for accurate gating of different cell populations during analysis.)
5. To analyze the flow data, with a flow cytometry analysis program, select the correct cell population to analyze. For two-step gating, use forward scatter (FS) and side scatter (SS) to discriminate the cells based on size and granularity, respectively. Doublets and aggregates have a higher time of flight, and they can be eliminated at this step.
6. Since dead cells bind nonspecifically to many reagents and antibodies, they could lead to false signals in flow cytometry. Therefore, use the data from the tube containing cells stained solely with the viability marker to gate out dead cells. Now that viable populations are selected, gate the desired cell populations using the FMO samples. As the Rodeheffer group

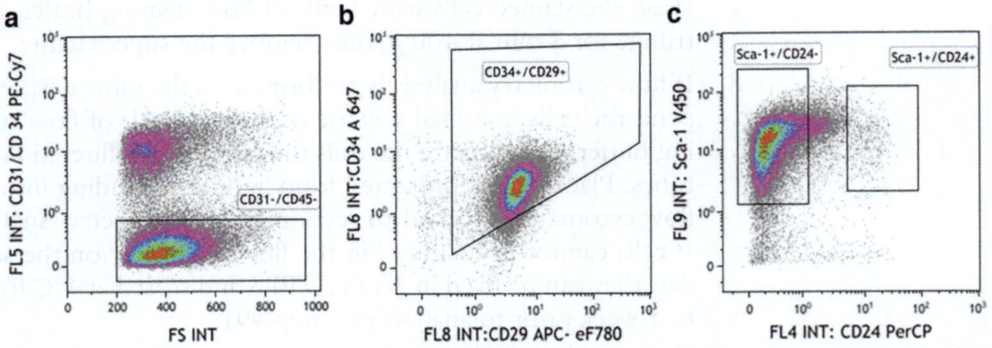

Fig. 2 Flow cytometry plots identifying adipocyte progenitor cells. (**a**) CD31⁻ and CD45⁻ population, using antibodies conjugated with PE-Cy7 fluorochrome. (**b**) CD34⁺ and CD29⁺ population, using antibodies conjugated with Alexa Fluor 647 and APC-eFluor 780 fluorochromes, respectively. (**c**) Sca-1⁺ and CD24⁺ and CD24⁻ populations, using antibodies conjugated with V450 and PerCP fluorochromes, respectively

demonstrated [23], first gate the lineage-negative cells, which are defined as $CD31^-$ and $CD45^-$. Next, the selected population will be gated with $CD29^+$ and $CD34^+$. Subsequently, the double-positive population will be gated for $Sca1^+$ and $CD24^+$ or $CD24^-$ population (Fig. 2a–c). APCs can also be identified through the expression of Pdgfrα^+.

Fluorescence-activated Cell Sorting (FACS)

FACS enables the isolation of unique cell populations based on cell surface profile, and it serves as a useful tool for purifying APCs. The general protocol for FACS is similar to flow cytometry. However, depending on the intended downstream applications in which the cells will be used, additional caution should be taken to use sterile reagents, when possible, particularly if cells are cultured. Similar control tubes are prepared for FACS. Prepare sterile Eppendorf® tubes containing collecting medium and sort the desired number of respective cell populations into the prepared tubes. As shown in Fig. 3, both endogenous tissue fluorescent markers and antibodies can be used to identify and sort distinct APC populations.

1. After cell sorting, centrifuge cells at 300 *g* for 5 min and resuspend in 100–500 μL plating medium depending on the number of sorted cells. Approximately 10,000–20,000 cells are seeded in each well in a 96-well plate (*see* **Note 6**).
2. Swirl cells for even distribution and incubate at 37°C. The seeded cells should not be disturbed for 24–48 h.
3. Replace media after 48 h with DMEM/F12 containing 10% FBS.
4. The confluent cells can be differentiated into adipocytes using the adipogenic differentiation media. After 2 days, the media should be changed to DMEM/F12 containing 10% FBS with

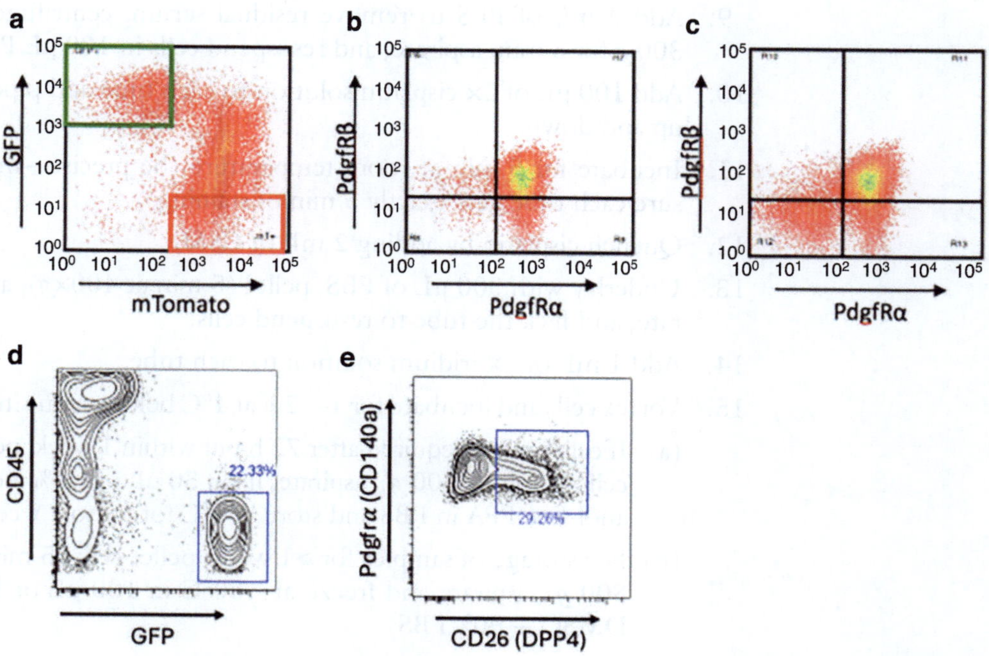

Fig. 3 Identification of APC populations based on fluorescent reporters and expression of Pdgfra. (**a**) SVF was analyzed by GFP (mG) and Tomato (mT). (**b**) mG population analyzed by PDGFR-α and PDGFR-β using PE-Cy7 and APC fluorochrome, respectively. (**c**) mT population analyzed by PDGFR-α and PDGFR-β using PE-Cy7 and APC fluorochrome, respectively. (**d**) Pdgfrα⁺ APCs identified based on GFP expression in Pdgfrα-eGFP mice. (**e**) DPP4-expressing APCs gated on Pdgfrα (CD140a) and CD26 (DPP4) using PE and BV421 fluorochromes, respectively

Insulin (1 μM). Lipid droplet-filled adipocytes will be observed after 4 days of differentiation.

3.5 Preparation of Cells for CyTOF Analysis

1. Aliquot 1–3 million cells into 4 mL conical tubes (*see* **Note 7**).
2. Add 1 mL of CSM, then underlay with 300 μL of FBS.
3. Pellet the cells by centrifugation at 300 *g* for 5 min and carefully aspirate the supernatant.
4. Optional barcoding step: If the sample cell number is low or the total number of samples is high, we strongly recommend barcoding individual samples and pooling into 1 sample prior to immunostaining (*see* **Note 8**).
5. Suspend the cells in 25 μL of FcR blocking solution and incubate at room temperature for 10–15 min (or 45 min on ice).
6. Add 25 μL of 2× surface Ab cocktail (for a final volume of 50 μL of 1× antibody staining mix) and pipette a few times to ensure adequate mixing.
7. Incubate for 30 min at room temperature.
8. Wash with 1 mL of CSM, pellet (5 min at 300×*g*), aspirate, and flick the tube.

9. Add 2 mL of PBS to remove residual serum, centrifuge at 300 *g* for 5 min, aspirate, and resuspend cells in 100 μL PBS.
10. Add 100 μL of 2× cisplatin solution to each tube and pipette up and down.
11. Incubate for 5 min at room temperature (be precise—make sure each tube gets exactly 5 min of staining).
12. Quench cisplatin by adding 2 mL of CSM.
13. Underlay with 300 μL of FBS, pellet (5 min at 400×*g*), aspirate, and flick the tube to resuspend cells.
14. Add 1 mL of 1× iridium solution to each tube.
15. Vortex cells and incubate for 1–72 h at 4°C before acquisition.
    (a) If cells will be acquired after 72 h but within 1 week, pellet cells (5 min at 800 *g*), aspirate, fix in 50 μL of 1.6% methanol-free PFA in PBS and store at 4°C for up to 1 week.
    (b) For storage of samples for >1 week, pellet cells (5 min at 800 *g*), aspirate and freeze at −80°C in 100 μL of 10% DMSO + 90% FBS).
16. About 1 h before acquisition on the Helios mass cytometer, pellet (5 min at 800×*g*), aspirate, and flick.
17. Add 3 mL of CSM, pellet (5 min at 800×*g*), aspirate, and flick.
18. Suspend in 1 mL of PBS and take an aliquot to count cells. The cell count is needed to determine the volume of normalization beads added to samples prior to sample acquisition. Wash cells in 2–3 mL of PBS, centrifuge for 5 min at 800 *g*, aspirate and store pellets on ice prior to acquisition.
19. The samples will be mixed with normalization beads by the CyTOF operator within the facility and will be acquired on a Helios mass cytometer.

3.6 CyTOF Analysis

Here, we briefly describe steps for CyTOF data analysis using the CytoBank software platform. There are currently other software options that can also be used for the analysis of CyTOF data. Gating will be conducted to identify live single cells from dead cells, doublets, and noncellular debris.

1. Use biaxial gating of iridium (DNA-Ir191), A DNA intercalator vs EQ normalization beads (EQ-140Ce), which are used as a control to normalize for variability in instrument signal detection over time [31]. Draw a gate around the iridium DNA-positive and EQ-negative region, which can be named "DNA+ Cells."
2. Next, select the newly formed "DNA+ Cells" population. Create a second biaxial plot of Event Length vs Center, a Gaussian parameter, to isolate single cells from doublets or cellular aggregates. Use the polygon selection tool to select the dense cluster

or region positive for Center and Event Length, while excluding any sparse peripheral events. Name the gate "EL vs Center." Ensure that the EL parameter scale is visualized on a log scale.

3. Select the EL vs Center population and create a biaxial plot of Event Length vs Offset, another Gaussian parameter, to further refine the single-cell events and exclude any noise. Use the polygon selection tool to select the most densely populated region of events and name the gate "EL vs Offset."
4. Select "EL vs Offset under population. The final quality control gate is to isolate viable single cells via Cisplatin staining. Plot Cisplatin vs any cellular marker of choice on a biaxial plot and create a gate around the region negative for Cisplatin, which represents live singlets. Name the gate accordingly.
5. Based on the panel of antibodies used for staining, perform gating and comparisons for specific target cell types. Similar to flow cytometry analysis, adipose progenitor cell surface staining markers (e.g., CD45−, CD31−, $CD34^+$, $Pdgfr\alpha^+$, CD24, Sca-1) can be used to visualize the APC population. Furthermore, markers of heterogeneous APC populations such as CD54 (ICAM-1), CD24 (DPP4), CD142, and CD9 can be included to identify and characterize adipogenic, stem-like progenitor cells, and fibrogenic cells as previously described [11].
6. Use advanced analyses, including tSNE, UMAP, viSNE, and FlowSOM to enable analysis and visualization of APCs and simultaneous immunophenotyping of adipose immune cell populations [11] (Fig. 4).

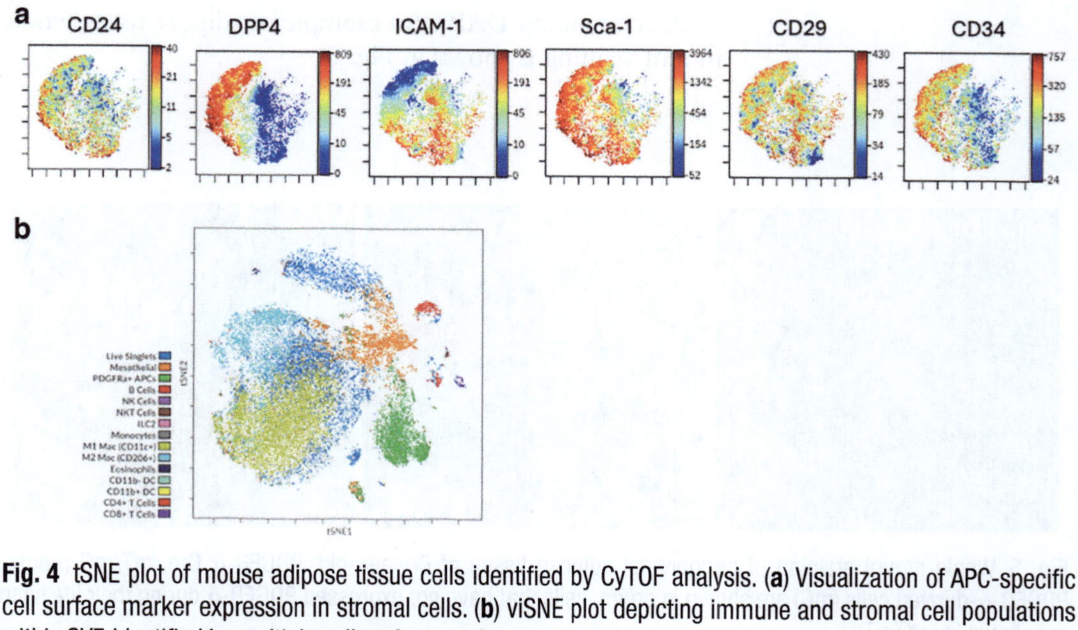

Fig. 4 tSNE plot of mouse adipose tissue cells identified by CyTOF analysis. (**a**) Visualization of APC-specific cell surface marker expression in stromal cells. (**b**) viSNE plot depicting immune and stromal cell populations within SVF, identified by multiple cell surface markers

3.7 Whole-Mount Staining of Adipose Tissue and Visualization

Adipose tissue can be observed under the microscope without sectioning, using a technique called whole-mount staining. This method is mostly used for adipose depots with fewer cells and lower vessel density, such as perigonadal adipose tissue. To prepare for whole-mount staining of adipose tissue:

1. Dissect the adipose tissue with blunt-ended tools to avoid stretching or damaging the tissue.
2. Rinse the tissue in fixation media overnight at 4°C or 1–2 h at RT.
3. Wash the tissue on ice using cold PBS for 10 min, using a shaker. Repeat this step 3 times.
4. If using a multicolor reporter mouse line, such as mT/mG , the tissue is ready for microscopy. However, if further antibody staining is required, follow the next steps.
5. Permeabilize and block the tissue using the blocking medium for 1 h.
6. Incubate with primary antibodies, diluted in blocking solution, overnight at 4°C.
7. The next day, wash the tissue on ice using cold PBS for 10 min, using a shaker. Repeat this step 3 times.
8. Incubate with fluorescence-conjugated secondary antibodies diluted in PBS with 1% BSA for 1 h at room temperature.
9. Wash the tissue as explained in step #3.
10. The tissue is ready for microscopy. Apply nonsolidifying mounting media and visualize the tissue using a confocal microscope. If nuclear staining is required, use mounting media containing DAPI. An example of adipose tissue whole mount staining is shown in Fig. 5.

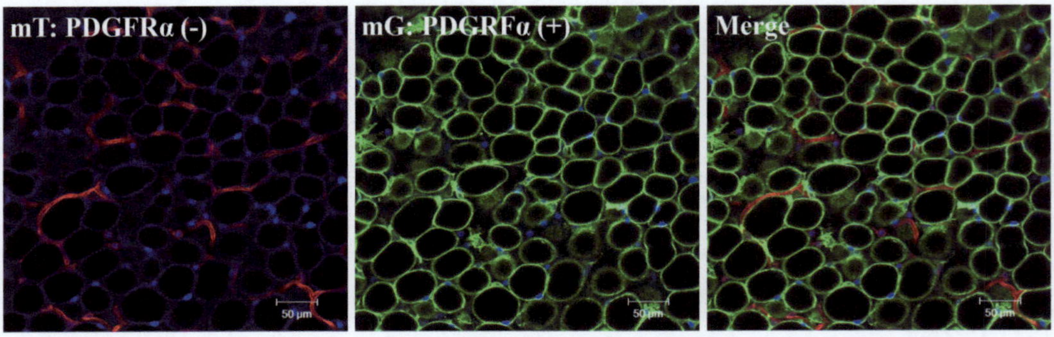

Fig. 5 Whole mount staining of perigonadal adipose tissue of 8-week-old, PDGFR-α Cre; mT/mG mouse. PDGFR-α-derived cells (mG) are shown in green; cells that have not expressed PDGFR-α during their life span are shown in red (mT)

4 Notes

1. While this is optional, we recommend storage of tissues in Miltenyi storage solution at 4°C if tissues cannot be digested within a reasonable amount of time, to increase cell yield and viability.
2. Avoid overdigestion (>3 min of chopping) of tissue, as it can increase cell death. Dissect out the lymph node from the inguinal or mesenteric adipose depot if needed.
3. Before washing the unfiltered tissue, gently mash the undigested tissue pieces with a blunt tool and wash with extra wash-ing buffer to increase cell yield.
4. Depending on the number of adipose tissues to be digested and analyzed, it is often not possible to perform same-day live cell flow cytometry analysis. Therefore, we recommend using a fixable viability stain to ensure accurate assessment of cell viability in samples that will be fixed and stored for analysis on a separate day. We routinely stain cells with the viability dye prior to staining with the cell surface antibody mix at a concentration of 1:500–1:1000 for 20 min. However, follow the manufacturer's protocol to determine the optimum staining conditions. In the case of live cell same-day analysis or FACS, cells can be stained with a nonfixable viability stain such as DAPI or propidium iodide (PI), which can be added to the cells just prior to acquisition on the flow cytometer.
5. When preparing cells for antibody staining, pool excess cells from samples with larger cell pellets to prepare for FMO samples.
6. To minimize cell death, it is ideal to seed the sorted cells as soon as possible.
7. In comparison to cell preparation for flow cytometry analysis, CyTOF cell preparation leads to more cell loss. Expect a loss of ~30–50% of cells throughout the staining process. To minimize cell loss, be careful with aspiration of cells after washes and use conical tubes to better facilitate visualization of the cell pellet.
8. The Cell-ID 20-Plex Pd Barcoding Kit (Standard BioTools) allows for unique barcoding of up to 20 samples, enabling them to subsequently be stained and acquired as one multiplexed sample. We have performed barcoding when the starting cell number per sample is low. While we have not included the barcoding method within this protocol, it can be performed prior to Step 5 according to the manufacturer's protocol.

Acknowledgments

We thank the Center for Advanced Single Cell Analysis at The Hospital for Sick Children Research Institute for their considerable technical contribution, including optimization and troubleshooting of the CyTOF protocol and antibody panel design. This work was funded by grants from the Canadian Institute of Health Research (CIHR, PJT-190016 and PJT-162083), the Natural Sciences and Engineering Research Council of Canada (NSERC, RGPIN-2016-06610), and the Pilot and Feasibility Study Grant of Banting & Best Diabetes Centre (BBDC) to H.-K.S. J.H.L. is supported by the Restracomp scholarship from The Hospital for Sick Children. Q.W.Z. is supported by the Chinese Scholarship Council. N.V. is supported by Novo-Nordisk Studentship from BBDC and Restracomp Master's Scholarship from The Hospital for Sick Children.

Conflict of Interest The authors have no conflicts of interest to declare that are relevant to the content of this chapter.

References

1. Park A, Kim WK, Bae KH (2014) Distinction of white, beige and brown adipocytes derived from mesenchymal stem cells. World J Stem Cell 6(1):33–42
2. Zwick RK, Guerrero-Juarez CF, Horsley V, Plikus MV (2018) Anatomical, physiological, and functional diversity of adipose tissue. Cell Metab 27(1):68–83
3. Wang QA, Tao C, Gupta RK, Scherer PE (2013) Tracking adipogenesis during white adipose tissue development, expansion and regeneration. Nat Med 19(10):1338–1344
4. Wang W, Seale P (2016) Control of brown and beige fat development. Nat Rev Mol Cell Biol 17(11):691–702
5. Harms M, Seale P (2013) Brown and beige fat: development, function and therapeutic potential. Nat Med 19(10):1252–1263
6. Jung SM, Sanchez-Gurmaches J, Guertin DA (2019) Brown adipose tissue development and metabolism. Handb Exp Pharmacol 251:3–36
7. Uhm M, Saltiel AR (2015) White, brown, and beige; type 2 immunity gets hot. Immunity 42(1):15–17
8. Boström P, Wu J, Jedrychowski MP, Korde A, Ye L, Lo JC, Rasbach KA, Boström EA, Choi JH, Long JZ, Kajimura S, Zingaretti MC, Vind BF, Tu H, Cinti S, Højlund K, Gygi SP, Spiegelman BM (2012) A PGC1-alpha-dependent myokine that drives brown-fat-like development of white fat and thermogenesis. Nature 481(7382):463–468
9. Xu X, Ying Z, Cai M, Xu Z, Li Y, Jiang SY, Tzan K, Wang A, Parthasarathy S, He G, Rajagopalan S, Sun Q (2011) Exercise ameliorates high-fat diet-induced metabolic and vascular dysfunction, and increases adipocyte progenitor cell population in brown adipose tissue. Am J Physiol Regul Integr Comp Physiol 300(5):R1115–R1125
10. Ziqubu K, Dludla PV, Mthembu SXH, Nkambule BB, Mabhida SE, Jack BU, Nyambuya TM, Mazibuko-Mbeje SE (2023) An insight into brown/beige adipose tissue whitening, a metabolic complication of obesity with the multifactorial origin. Front Endocrinol (Lausanne) 14:1114767
11. Lee JH, Ealey KN, Patel Y, Verma N, Thakkar N, Park SY, Kim JR, Sung HK (2022) Characterization of adipose depot-specific stromal cell populations by single-cell mass cytometry. iScience 25(4):104166
12. Iyer A, Hamers AAJ, Pillai AB (2022) CyTOF (®) for the masses. Front Immunol 13:815828
13. Silva Ribeiro DJ, Yüksel S, Dolf A, Wachten D (2024) Isolation and flow cytometry analysis of macrophages from white adipose tissue. Methods Mol Biol 2713:149–158

14. Rondini EA, Granneman JG (2020) Single cell approaches to address adipose tissue stromal cell heterogeneity. Biochem J 477(3):583–600
15. Feil S, Valtcheva N, Feil R (2009) Inducible Cre mice. Methods Mol Biol 530:343–463
16. Jeffery E, Berry R, Church CD, Yu S, Shook BA, Horsley V, Rosen ED, Rodeheffer MS (2014) Characterization of Cre recombinase models for the study of adipose tissue. Adipocyte 3(3):206–211
17. Krueger KC, Costa MJ, Du H, Feldman BJ (2014) Characterization of Cre recombinase activity for in vivo targeting of adipocyte precursor cells. Stem Cell Rep 3(6):1147–1158
18. Berry DC, Jiang Y, Graff JM (2016) Mouse strains to study cold-inducible beige progenitors and beige adipocyte formation and function. Nat Commun 7: 10184
19. Kretzschmar K, Watt FM (2012) Lineage tracing. Cell 148(1-2):33–45
20. Livet J, Weissman TA, Kang H, Draft RW, Lu J, Bennis RA, Sanes JR, Lichtman JW (2007) Transgenic strategies for combinatorial expression of fluorescent proteins in the nervous system. Nature 450(7166):56–62
21. Berry R, Rodeheffer MS (2013) Characterization of the adipocyte cellular lineage in vivo. Nat Cell Biol 15(3):302–308
22. Muzumdar MD, Tasic B, Miyamichi K, Li L, Luo L (2007) A global double-fluorescent Cre reporter mouse. Genesis 45(9):593–605
23. Rodeheffer MS, Birsoy K, Friedman JM (2008) Identification of white adipocyte progenitor cells in vivo. Cell 135(2):240–249
24. Merrick D, Sakers A, Irgebay Z, Okada C, Calvert C, Morley MP, Percec I, Seale P (2019) Identification of a mesenchymal progenitor cell hierarchy in adipose tissue. Science 364(6438):eaav2501
25. Emont MP, Jacobs C, Essene AL, Pant D, Tenen D et al (2022) A single-cell atlas of human and mouse white adipose tissue. Nature 603(7903):926–933
26. Marcelin G, Ferreira A, Liu Y, Atlan M, Aron-Wisnewsky J, Pelloux V, Botbol Y, Ambrosini M, Fradet M, Rouault C, Hénégar C, Hulot JS, Poitou C, Torcivia A, Nail-Barthelemy R, Bichet JC, Gautier EL, Clément K (2017) A PDGFRα -Mediated switch toward CD9(high) adipocyte progenitors controls obesity-induced adipose tissue fibrosis. Cell Metab 25(3):673–685
27. Tang W, Zeve D, Suh JM, Bosnakovski D, Kyba M, Hammer RE, Tallquist MD, Graff JM (2008) White fat progenitor cells reside in the adipose vasculature. Science 322(5901):583–586
28. Cao Y (2010) Adipose tissue angiogenesis as a therapeutic target for obesity and metabolic diseases. Nat Rev Drug Discov 9(2):107–115
29. Tran KV, Gealekman O, Frontini A, Zingaretti MC, Morroni M, Giordano A, Smorlesi A, Perugini J, De Matteis R, Sbarbati A, Corvera S, Cinti S (2012) The vascular endothelium of the adipose tissue gives rise to both white and brown fat cells. Cell Metab 15(2):222–229
30. Gupta RK, Mepani RJ, Kleiner S, Lo JC, Khandekar MJ, Cohen P, Frontini A, Bhowmick DC, Ye L, Cinti S, Spiegelman BM (2012) Zfp423 expression identifies committed preadipocytes and localizes to adipose endothelial and perivascular cells. Cell Metab 15(2):230–239
31. Bagwell CB, Inokuma M, Hunsberger B, Herbert D, Bray C, Hill B, Stelzer G, Li S, Kollipara A, Ornatsky O, Baranov V (2020) Automated data cleanup for mass cytometry. Cytometry A 97(2):184–198

Chapter 16

Investigating Protein Interactions in Single Cells by Proximity Ligation Assay (PLA)

Zoe E. Gillespie, Saloni Modi, Mariia Cherednychenko, Mario Filice, Navroop K. Dhaliwal and Jennifer A. Mitchell

Abstract

Many cellular processes are governed by protein–protein interactions. To better understand these processes, it is necessary to understand how different proteins interact under different conditions. Proximity ligation assay (PLA) enables identification of proteins within 40 nm of one another, suggesting interaction, while also providing information on sub-cellular localization of the interaction. This information is particularly informative when monitoring the localization and interaction of transcription factors and other nuclear proteins. Here, we describe an optimized protocol for PLA in mouse embryonic stem cells, trophoblast stem cells, and blastocysts, alongside a demonstration of a robust positive control for nuclear proteins.

Keywords Proximity ligation assay, Protein–protein interaction, Mouse, Bastocyst, Embryonic stem cells, Trophoblast stem cells, Immunolabeling

1 Introduction

Protein–protein interaction is fundamental for cellular function and can indicate how cells respond to developmental cues. Proximity ligation assay (PLA) is a technique that detects and visualizes protein co-association within unmodified tissue and cell samples [1]. In a PLA, two different primary antibodies, raised in separate species, identify two distinct protein targets [2, 3]. Subsequently, secondary antibodies labeled with oligonucleotides (PLA probes) detect the primary antibodies [2, 3]. If the proteins of interest are within 40 nm of each other, the DNA probes hybridize to form a circular DNA template necessary for rolling-circle amplification (RCA) [2]. Finally, fluorescently-labeled oligos bind to complementary sequences within the amplicon, generating foci that can be quantified through microscopy image analysis [2, 3]. This process is graphically represented in (Fig. 1). By combining specific target binding with amplification and signal detection steps, PLA achieves high sensitivity and specificity in

Paul Delgado-Olguín (ed.), *Mouse Embryogenesis*, Methods in Molecular Biology, vol. 3010,
https://doi.org/10.1007/978-1-0716-5126-1_16,

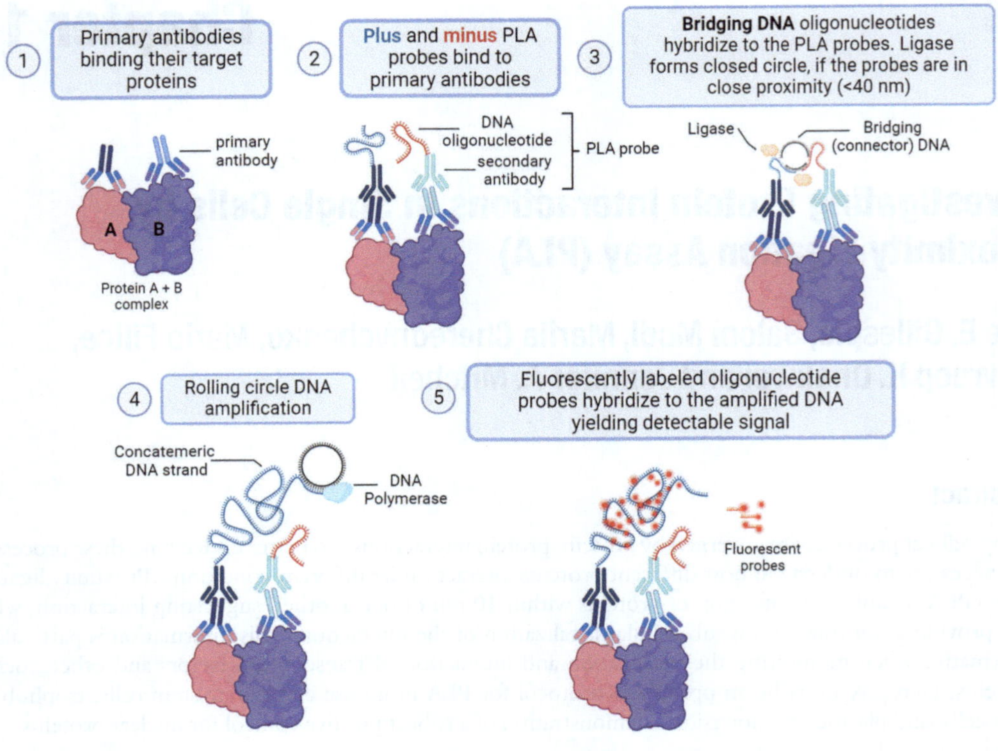

Fig. 1 The PLA reaction. Permeabilized cells are incubated with primary antibodies, followed by the addition of PLA probes consisting of DNA oligonucleotide fragments (plus and minus) attached to secondary antibodies. If the target proteins of the primary antibodies are within 40 nm of each other, the probes ligate with the help of bridging oligonucleotides and DNA ligase, forming a template for DNA polymerase. This initiates rolling circle DNA amplification, producing a long concatemer attached to the secondary antibody, and hence to the protein complex. Fluorescently labeled complementary oligonucleotide probes then hybridize to the amplicon, generating a signal that can be visualized by fluorescent confocal microscopy. Created with BioRender.com.

detecting and quantifying target protein–protein interactions, while localizing these interactions to subcellular compartments [1–3].

PLA offers advantages over immunoprecipitation by allowing the analysis of individual proteins, protein–protein interaction, or specific protein modifications, along with observing subcellular localization [2, 4, 5]. This is especially useful when genetic manipulation is challenging, such as when analyzing paraffin-embedded or frozen tissue sections or whole-mount pre-implantation embryos. One limitation of PLA is its dependence on the availability and specificity of primary antibodies. Validating the specificity of primary antibodies before starting a PLA experiment is crucial for success. The positive control antibodies suggested below both detect the large subunit of RNA polymerase II, RPB1. One antibody recognizes the C-terminal domain of the protein when phosphorylated on serine 5 of the repeat domain, and the other antibody recognizes a domain in the core of the protein. This combination should provide a positive signal in most cells as these domains are highly conserved, and all cells express RPB1. Here, we present an optimized PLA protocol and positive control reaction for mouse embryonic stem (ES) cells, trophoblast stem cells and blastocysts, based on investigating interactions between transcription factors and signalling pathways [6, 7] (Fig. 2).

2 Materials

2.1 Coverglass Preparation

1. Circular cover glass (18 mm diameter, #1 ½ thickness).
2. Clear 12-well multiwell plates (flat bottom, tissue-culture treated polystyrene).
3. 70% ethanol (mix 35 mL 100% ethanol with 15 mL MilliQ water).

2.2 Immunolabeling

1. 10% neutral buffered formalin.
2. 1× phosphate-buffered saline (PBS)
3. Block/permeabilization buffer for mouse embryonic stem cells and trophoblast stem cells: 10% fetal bovine serum, 0.1% Triton X-100, in 1× PBS (*see* **Note 1**).
4. Permeabilization buffer for mouse blastocysts: 0.3% Triton X-100 in 1× PBS.
5. Block buffer for mouse blastocysts: 10% goat serum, 0.1% Triton X-100 in 1× PBS
6. Antibody diluent for mouse embryonic stem cells and trophoblast stem cells: Duolink® In Situ Detection Reagents FarRed or 0.2% fetal bovine serum, 0.1% Triton X-100 in 1× PBS.
7. Antibody diluent for mouse blastocysts: 5% goat serum, 0.1% Triton X-100 in 1× PBS
8. Rabbit anti-RNA polymerase II CTD repeat YSPTSPS (phospho S5).
9. Mouse anti-RNA polymerase II, clone ARNA-3.
10. Immunolabeling wash buffer: 0.1% Tween 20, 1× PBS.

2.3 Proximity Ligation Assay

1. Duolink® in situ PLA® probe anti-rabbit PLUS (*see* **Note 2**).
2. Duolink® in situ PLA® probe anti-mouse MINUS (*see* **Note 2**).
3. Duolink® in situ detection reagents FarRed (*see* **Note 3**).
4. Duolink® in situ mounting medium with DAPI.
5. Duolink® in situ wash buffers, fluorescence.
   a. 1× wash buffer A. Make a purchased buffer up to 1 L with MilliQ water and store at 4°C or weigh 8.8 g NaCl, 1.2 g Tris base and 0.5 mL Tween20 and dissolve in 800 mL

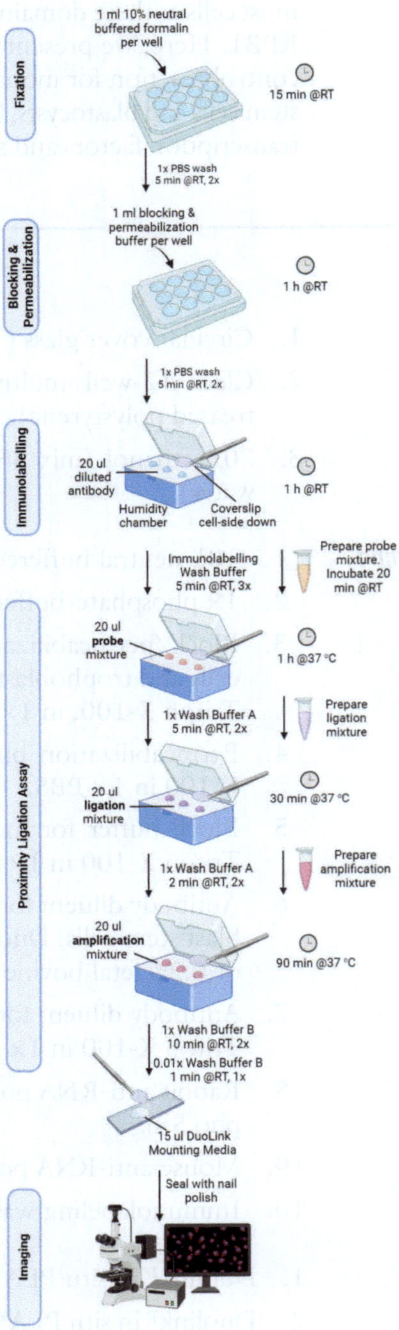

Fig. 2 PLA experimental workflow. Overview of the proximity ligation assay workflow. For a more detailed protocol, see the text. Note that each step has specific incubation conditions and is followed by washes with different buffers and repetitions. Cells cultured on coverslips are fixed with 10% neutral buffered formalin, followed by permeabilization and blocking. For immunolabeling, the coverslips are transferred to a humidity chamber (a box lined with parafilm on top of PBS-soaked paper). Primary antibody droplets (20 μL) are placed on the parafilm, and the coverslips are placed cell-side down on the

high purity water; using HCl, adjust pH to 7.4 and bring the solution to a final volume of 1 L. Optional: filter through a 0.22 μm filter.

b. 1× wash buffer B. Make purchased buffer up to 1 L with MilliQ water and store at 4°C or weigh 5.84 g NaCl, 4.24 g Tris base, and 26.0 g Tris HCl and dissolve in 800 mL high purity water; using HCl, adjust pH to 7.5 using HCl and bring solution to a final volume of 1 L. Optional: filter through a 0.22 μm filter.

6. Molecular biology grade Sigma water.

2.4 Blastocyst Preparation

1. Human chorionic gonadotropin (hCG) stock (50 IU); hCG is reconstituted from a lyophilized powder at an activity of 10,000 IU per vial, and 50 μL aliquots (50 IU per aliquot) are made and stored at −80°C until needed. Each tube can be thawed and diluted with 950 μL of PBS and is sufficient to provide enough hormone for 10 females (0.1 mL or 5 IU per mouse).
2. Pregnant mare serum gonadotropin (PMSG) stock (50 IU). PMSG is reconstituted from a lyophilized powder at an activity of 1000 IU per vial, and 50 μL aliquots (50 IU per aliquot) are made and stored at −80°C until needed. Each tube can be thawed and diluted with 950 μl of PBS and is sufficient to provide enough hormone for 10 females (0.1 mL or 5 IU per mouse).
3. Sterile 1× PBS.
4. M2 medium.
5. RainX, glass cleaner and rain repellent.
6. 4-well glass dish for embryo transfer, pre-coated with RainX to prevent sticking of the embryos to the glass. Or use 4-well dishes for IVF, which do not need additional coating.
7. 96-well dish.

2.5 Equipment

1. Incubator set to 37°C: for probe, ligation, and amplification incubation steps.
2. Humidity chamber.
3. Fluorescence/confocal microscope with appropriate channels, camera, and software for image acquisition.

Fig. 2 (continued) droplets. This process is repeated with fresh parafilm and 20 μL droplets of probes, ligation, and amplification mixtures consecutively. Following the final wash, dried coverslips are mounted on glass slides with DuoLink Mounting Media, cell-side down, and sealed with nail polish. The slides are then ready for imaging by fluorescent confocal microscopy. Created with BioRender.com.

4. Tweezers for coverslips.
5. Tweezers for blastocyst preparation.
6. Mouth pipetting equipment for blastocyst transfer.
7. Pasteur pipettes, borosilicate glass, long tip.
8. Bunsen burner.
9. 1 mL syringe.
10. 26.5-guage needle

2.6 Analysis

Imaris Image Analysis Software: Oxford Instruments; or ImageJ: https://imagej.net/ij/.

3 Methods

3.1 Preparation of Coverslips for PLA in Mouse Embryonic Stem Cells and Trophoblast Stem Cells

1. Prepare coverslips for use in cell culture by sterilizing in 70% ethanol in a biosafety cabinet. Ensure all ethanol has evaporated from the coverslip before use.
2. Add a coverslip to the well/growth plate and prepare the well as usual.
3. After adding cells and media to the well, check that no bubbles are present underneath the coverslip and that the coverslip is in contact with the bottom of the culture dish. This ensures cell growth only on the top-facing and exposed surface of the coverslip.

3.2 Adherent Cell Fixation, Blocking, Permeabilization, and Storage

Since fixation and permeabilization can vary between cell lines and tissues, begin optimization of this process for PLA by fixing, permeabilizing, and blocking as done for immunofluorescence in your cell or tissue. The described protocol and reagents used have been optimized in mouse embryonic stem cells and trophoblasts.

1. Once cells have reached the desired confluence, aspirate the growth media.
2. In 12-well dishes, add 1 mL of 10% neutral buffered formalin and incubate for 15 min at RT.
3. Aspirate neutral buffered formalin and wash twice with 1 mL/well 1× PBS. Samples can be stored for 1–2 weeks at 4°C in 1× PBS at this point.
4. Add 1 mL block and permeabilization buffer to each well, incubate at RT for 1 h.
5. Aspirate block and permeabilization buffer and wash twice with 1 mL 1× PBS, 5 min each.

3.3 Immunolabeling

Reliable PLA results depend on the appropriate selection and optimization of primary antibodies to reduce background and improve specificity (*see* **Note 4**).

1. If stored at 4°C, bring samples to RT.
2. Prepare the humidity chamber with filter paper and parafilm.
3. Dilute rabbit anti-RNA polymerase II CTD 1:1000 and mouse anti-RNA polymerase II (core) 1:2000 in antibody diluent provided by Duolink® in situ detection reagents kit or 0.2% fetal bovine serum, 0.1% Triton X-100 in 1× PBS.
4. Apply 20 μL of antibody dilution as a droplet (*see* **Note 5**) to the parafilm in the humidity chamber and place the coverslip cell-side down onto the antibody droplet.
5. Incubate for 1 h RT or 4°C overnight.

3.4 Proximity Ligation Assay of Adherent Embryonic Stem Cells/Trophoblast Stem Cells

Allow buffers to come to room temperature prior to use and prewarm a humidity chamber to 37°C.

1. Prepare probe mixture by combining minus probe (4 μL/sample), plus probe (4 μL/sample), and molecular biology grade water (12 μl/sample). Mix by pipetting and incubate at RT for 20 min.
2. While the probe is incubating, gently lift coverslips from parafilm inside the humidity chamber and transfer to immunolabeling wash buffer. Incubate for 5 min at RT. Repeat this wash three times.
3. Place 20 μL droplets of probe mixture onto new parafilm in the humidity chamber.
4. Remove excess moisture from the coverslip to avoid dilution of the probe and increased background noise. Hold the coverslip with forceps at an angle to accumulate a liquid droplet by gravity before dabbing the corner of the slide on absorbent tissue/Kim wipe.
5. Place the coverslip cell-side down onto the probe by gently lowering the coverslip with forceps, ensuring no bubbles or sliding of the coverslip.
6. Incubate for 1 h at 37°C.
7. Wash twice in 1× wash buffer A at RT, 5 min/wash.
8. Prepare ligation mixture by adding molecular biology grade water (15.5 μL/sample), 5× ligation buffer (4 μL/sample), and ligase (0.5 μL/sample). Mix well and collect by brief centrifugation.
9. Prepare the humidity chamber with new parafilm and add droplets of 20 μL of the ligation mixture.

10. Wick away excess moisture from the coverslip, place the cell-side down onto the ligation mixture droplet.
11. Incubate for 30 min at 37°C.
12. Wash twice in 1× wash buffer A at RT, 2 min/wash.
13. Prepare amplification mixture by combining molecular biology grade water (15.75 μL /sample), detection probe (4 μL/ sample) and polymerase (0.25 μL/sample).
14. Prepare the humidity chamber with new parafilm and add droplets of 20 μL of amplification mixture.
15. Wick moisture from coverslips and place them cell-side down onto the amplification mixture droplet. Incubate 90 min at 37°C.
16. Wash twice in 1× wash buffer B at RT, 10 min/wash.
17. Wash once in 0.01× wash buffer B at RT, 1 min.
18. To mount, add 15 μL of DuoLink mounting media to a glass microscope slide, remove the coverslip from the wash buffer, wick away excess moisture, and gently place the coverslip cell-side down on top of the mounting medium droplet. Seal with nail polish.
19. Samples can be imaged. Due to the structure of mouse stem cells and trophoblasts, if quantification of foci is required, confocal imaging should be conducted. Image analysis and foci quantification can be completed using Imaris [6, 7] or ImageJ software.

3.5 Proximity Ligation Assay in Mouse Blastocysts

Mouse Superovulation

This step is necessary to ensure a good number of blastocysts are retrieved prior to PLA. Females should be ~6 weeks of age at the time of PMSG administration and males 7–8 weeks (*see* **Note 6**). Males should also be confirmed to be fertile and caged individually prior to mating with a female.

1. Day 1: Scent exposure. Female mice are placed in the assigned male cage to allow a few moments of interaction between the pair. This will ideally induce estrus in female mice and enhance the chances of mating.
2. Day 1: PMSG injection. At 2 pm, inject the female intraperitoneally (IP) with 0.1 mL of freshly made PMSG in 1× PBS.
3. Day 2: No procedures scheduled.
4. Day 3: hCG injections. At noon (42–50 h post PMSG injection), inject each female with 0.1 mL of freshly made hCG in 1× PBS intraperitoneally (*see* **Note 7**).

5. Day 3: Mating. Housing both the female and male mouse together overnight (multiple females can be housed with a single male).
6. Day 4: Plug check. Early in the morning (7–8 am), check whether the females show a vaginal plug (evidence of mating). If so, separate the female from the male and count the days to e3.5, with the morning of the plug counted as e0.5. If the female does not show signs of a vaginal plug, leave the mice together for a further 24 h.
7. Day 5: Plug check. If the female did not show signs of a vaginal plug, check again early in the morning (7–8 am) for signs of a vaginal plug and separate from males if present.

Blastocyst Isolation

1. At e3.5 (3 days after a vaginal plug was seen), euthanize the mouse and check for eye or tail-pinch reflex to confirm.
2. Place the euthanized female on its back and spray the abdomen with ethanol. Cut open the abdominal skin to reveal the uterus. Ensure the oviduct remains, but the ovaries should be removed.
3. Dissect out the uterus into a 10 cm dish containing 5 mL of warmed M2 media.
4. Hold the uterine horns securely with tweezers and insert a 1 mL syringe containing M2 media into the end of the uterine horn and expel the media to flush the blastocysts from the uterus into the 10 cm dish, precoated with RainX to prevent embryos from sticking to the plastic.
5. Repeat for both sides of the uterus (this step can be repeated as many times as necessary to retrieve as many embryos as possible).
6. Wash blastocysts twice in 1×PBS using a mouth pipetting apparatus for blastocyst transfer between solutions. For more details on isolating and manipulating mouse embryos refer to Varghese et al. (2023) and Behringer et al. (2014) [8, 9].

3.6 Immunolabeling and PLA Protocol for Mouse Blastocysts

1. Stretch thin glass Pasteur pipettes' after heating them over the Bunsen burner flame from the middle of the tip in preparation for mouth pipetting the mouse blastocysts for transfer.
2. Transfer blastocysts to the 4-well glass dish, precoated with RainX containing 10% neutral buffered formalin. Fix at RT for 15 min.
3. Transfer blastocysts briefly to a well containing 1× PBS (*see* **Note 9**) before washing twice, 5 min/wash, in two additional wells containing 1× PBS at RT.
4. Transfer blastocysts briefly to 0.3% Triton X-100 in 1× PBS before permeabilizing in a second well of 0.3% Triton X-100 in 1× PBS for 15 min at RT.

5. Wash blastocysts twice in 1× PBS, 5 min/wash, RT.
6. Transfer blastocysts to 10% goat serum 0.1% Triton X-100 in 1× PBS briefly before blocking in a fresh volume of the same solution for 1 h at RT.
7. Dilute rabbit anti-RNA polymerase II CTD 1:1000 and mouse anti-RNA polymerase II (core) 1:2000 in 5% goat serum and 0.1% Triton X-100 in 1× PBS.
8. Prepare antibody dilution as a droplet in a well or on parafilm and transfer blastocysts to the droplet. Incubate for 1 h, RT.
9. Wash blastocysts in 0.05% Tween20 in 1× PBS three times, 5 min/wash, RT.
10. Repeat steps 1–17 of Sect. 3.4. For PLA reagents, we recommend the use of droplets for probe incubation, ligation, and amplification reactions. Blastocysts should be transferred between solution droplets at each step.
11. Blastocysts can be mounted either on glass microscope slides with concave regions, or on a slide to which nail polish has been used to section an area for the blastocyst to be mounted. The nail polish acts as a thin wall surrounding the blastocysts. This is necessary to avoid damaging or squishing the blastocyst when overlaying it with a coverslip.
12. Briefly incubate the blastocysts in a small droplet of DuoLink mounting medium before transferring to a new DuoLink droplet on the glass slide. Apply the coverslip and seal with nail polish.

4 Notes

1. FBS is derived from non-immunized animals and has no antigen specificity and is used to prevent (block) non-specific binding of antibodies to conserved sequences or FC-receptors, reducing non-specific/background signal. Goat serum serves the same function. Triton X-100 is a non-ionic detergent used to permeabilize the membrane of cells and allow antibody access to the antigen. Both blocking and permeabilization percentages and times may need to be optimized for different cell types and tissues or antibody combinations.
2. Antibody and probe selection: Primary antibodies must be raised against different species. Secondary antibodies (PLA probes) must be specific to the two species the primary antibodies were raised against and include a plus probe and a

minus probe. Here, we used mouse-anti and rabbit-anti primary antibodies. Therefore, secondary probes were a mouse minus and rabbit plus (but equally mouse plus and rabbit minus could be used).

3. The far red (Cy5) detection kit is used as it provides the best option for immunolabeling of different targets in other channels (Cy5 is often challenging in the detection of diffuse immunolabeled signals). Any detection wavelength is acceptable if it does not overlap with other secondaries used.
4. Experimental controls: Negative controls should include each primary antibody incubated independently with the PLA reaction (this will identify any background from nonspecific binding of secondary probes) and a third reaction in which no primary antibody is applied to samples, but the secondary antibodies (probe) reaction, ligation, and amplification occur (this should be negative for signal, or identify any baseline background from the PLA reagents). A positive control should include two primary antibodies that will produce foci (e.g., two primary antibodies targeting different regions of the same protein known to be within 40 nm of one another, such as the demonstrated RNA polymerase II interaction (Fig. 3)).
5. In total, 20 μL is the minimum volume of liquid we recommend using on a circular 18 mm diameter coverslip to prevent drying of the coverslip during 60 to 90 min 37°C incubations. This volume may need to be modified depending on the coverslip size.
6. Males under 6 months of age and that have a good breeding record are optimal, as older males may reduce the chances of a successful mating occurring.
7. Mouse response to hCG varies from strain to strain and dosage may vary depending on the strain used, the details provided here are recommended for C57/Bl6 mice. Generally, doses between 2.0 and 5.0 IU per mouse of each gonadotropin and females aged 21–42 days (body weights of 12–14 g) will give the best yields of fertilized eggs. Mice up to 8 weeks can be used, albeit with a lower yield of fertilized embryos.
8. The sensitivity of the assay in blastocysts may be increased through the removal of the Zona Pellucida by transferring embryos into 100–200 mL drops of room temperature acidic Tyrode's solution (Sigma-Aldrich T1788) until the zona pellucida is gone.

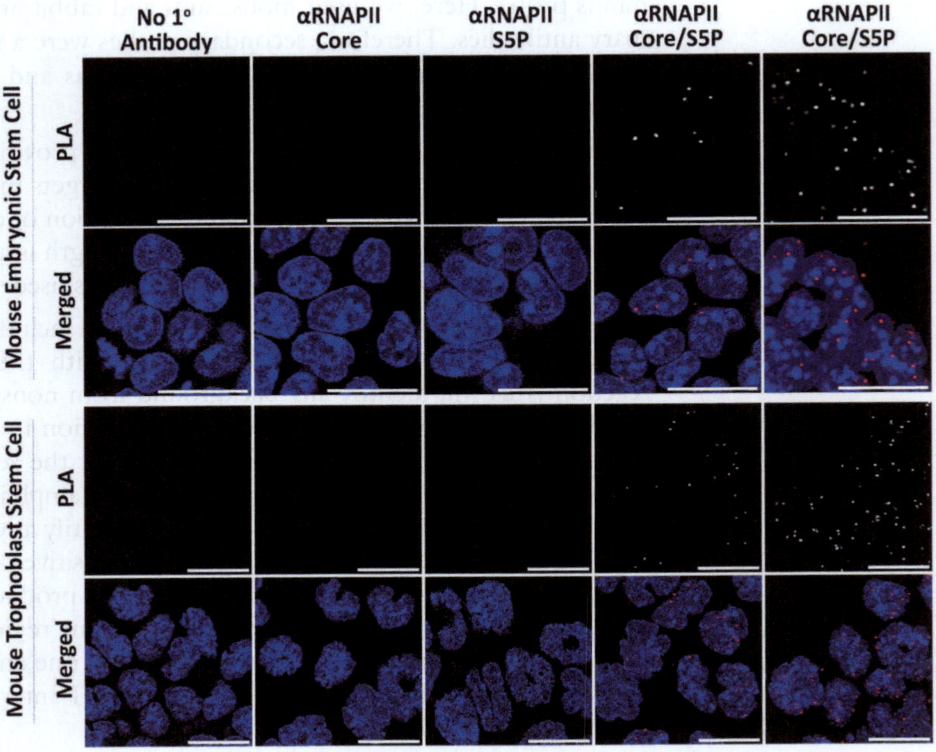

Fig. 3 Positive control reaction RNA polymerase II Core/S5P in mouse embryonic stem cells and mouse trophoblast stem cells. PLA demonstrating signal between two regions of the RNA polymerase II (RNAPII) protein in (top) F1 mouse embryonic stem cells, and (bottom) mouse trophoblast stem cells. From left to right, examples of no primary antibody control, RNA polymerase II Core and RNA polymerase II S5P only controls, and the combined PLA reaction with both RNA polymerase Core and S5P antibodies. The first four panels display a single optical section confocal image. The rightmost panel is a maximum intensity projection of the confocal image stack. Foci represent interaction between two regions of the RNA polymerase II protein (Core/S5P), a reliable positive control reaction due to the proximity of the regions. Greyscale images of foci are presented (PLA) with foci-nuclei merged overlays (Merged) for each cell line. In merged images, chromatin is counterstained with DAPI (blue), and foci are shown in red. Scale bar = 20 μm

9. For the transfer of blastocysts between wells/droplets of different solutions, we have included additional, but brief, wash steps to avoid dilution by any liquid transferred with the blastocyst.
10. If visualization of the entire pool of each protein is of interest, PLA can be combined with immunodetection by adding the secondaries for immunofluorescence after the PLA reaction is completed, followed by appropriate washes (Fig. 4). Care should be taken to avoid signal overlap between the PLA detection and the immunofluorescence secondary antibodies.

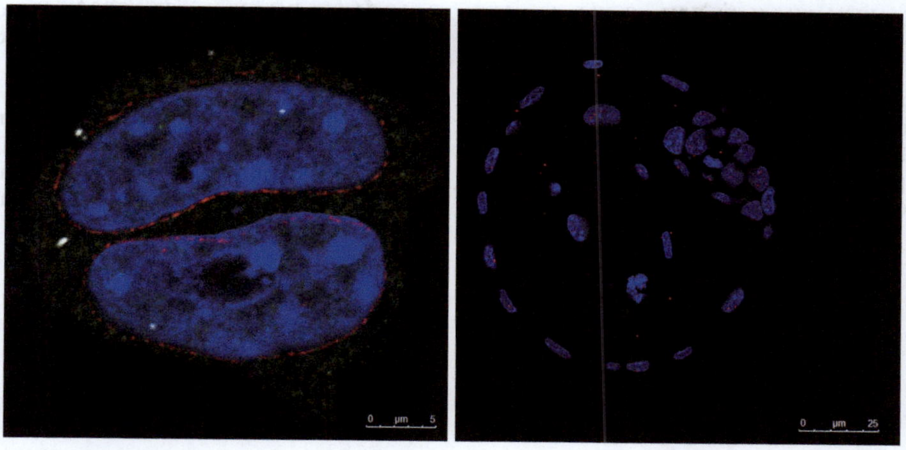

Fig. 4 Mouse blastocyst PLA, and PLA combined with immunodetection. Left: PLA combined with immunodetection in differentiating mouse embryonic stem cells. The nuclear export protein, XPO1, is labeled in red, KLF4, which is both nuclear and cytoplasmic in these cells, is labeled in green, and DAPI is shown in blue. The PLA signal is shown in white, which reveals the interaction between XPO1 and KLF4 as KLF4 is exported from the nucleus. Right: Mouse blastocyst PLA for KLF4 and XPO1 (red), DAPI shown in blue. The inner cell mass can be seen as the cluster of nuclei on the upper right of the image; the trophectoderm is the ring of cells surrounding the inner cell mass. The few nuclei in the center of the image are cells from the bottom layer of the trophectoderm. The scale is indicated on each image. Each image is a single optical section confocal image

Acknowledgments

This work was supported by the Canadian Institutes of Health Research (FRN PJT173252 and PJT180312), Natural Sciences and Engineering Research Council of Canada (2020-05972 discovery grant), the Canada Foundation for Innovation, and the Ontario Ministry of Research and Innovation (operating and infrastructure grants held by J.A.M.). Z.E.G. is funded by a CIHR postdoctoral fellowship.

Conflict of Interest The authors have no conflicts of interest to declare that are relevant to the content of this chapter.

References

1. Fredriksson S, Gullberg M, Jarvius J, Olsson C, Pietras K, Gustafsdottir SM, Ostman A, Landegren U (2002) Protein detection using proximity-dependent DNA ligation assays. Nat Biotechnol 20(5):473–477
2. Hegazy M, Cohen-Barak E, Koetsier JL, Najor NA, Arvanitis C, Sprecher E, Green KJ, Godsel LM (2020) Proximity ligation assay for detecting protein-protein interactions and protein modifications in cells and tissues in situ. Curr Protoc Cell Biol 89(1):e155
3. Alam MS (2018) Proximity ligation assay (PLA). Curr Protoc Immunol 123(1):e58
4. Lee C (2007) Coimmunoprecipitation assay. Methods Mol Biol 362:401–406
5. Ni D, Xu P, Gallagher S (2016) Immunoblotting and immunodetection. Curr Protoc Mol Biol 114:10.8.1–10.8.37

6. Dhaliwal NK, Miri K, Davidson S, El Jarkass HT, Mitchell JA (2018) KLF4 nuclear export requires ERK activation and initiates exit from naive pluripotency. Stem Cell Rep 10(4):1308–1323
7. Dhaliwal NK, Abatti L, Mitchell JA (2019) KLF4 protein stability regulated by interaction with pluripotency transcription factors overrides transcriptional control. Genes Dev 33(1):1–14
8. Varghese PC, Dutta D (2023) Protocol for isolation of mouse pre-implantation embryos for gene expression analysis. STAR Protoc 4(3):102479
9. Behringer R, Gertsenstein M, Vingersten Nagy K, Nagy S (2014) Manipulating the mouse embryo: a laboratory manual. Cold Spring Harbor Laboratory Press, New York

Check for updates

Chapter 17

Metabolic Plasticity in Embryogenesis Throughout the Lens of NAD^+

Quetzalcoatl Escalante-Covarrubias and Paul Delgado-Olguín

Abstract

Changes in cell fate during early mammalian development are supported by the dynamic regulation of energy metabolic pathways. These transitions fulfill energetic demands and influence transcriptional reprogramming and cell differentiation. Variations in NAD+/NADH ratios during early developmental transitions correlate with metabolic remodeling. Indeed, NAD+ bioavailability acts as an early signal for a change in the metabolic landscape by controlling the redox state and activity of multiple NAD+ dependent enzymes. Here, we summarize recent studies on NAD+ bioavailability and its function as the master organizer of energy metabolism during early mouse development. Additionally, we examine the function of NAD+ dependent chromatin remodelers in synchronizing transcriptional programs with pathways that fulfil the metabolic demands of developmental stage transitions.

Keywords NAD+/NADH radio, Redox state, Energy metabolism, Zygote, 2-, 4-cell stage, Blastocyst

1 Introduction

Oocyte fertilization initiates mammalian life, converts two differentiated germ cells into a totipotent zygote, and during pre-implantation development, the zygote forms into a blastocyst [1, 2]. These transitions are accompanied by extensive reprogramming of metabolic and epigenetic landscapes [3–5]. The Nicotinamide adenine dinucleotide (NAD+) is at the center of these changes, creating a dialog between metabolism and transcriptional programs expression of genes coordinating specific metabolic pathways during each stage of development.

The bioavailability of NAD+ is in a homeostatic status of biosynthesis, consumption and recycling [6, 7]. Mammalian cells can synthesize NAD+ de novo from tryptophan by the kynurenine pathway or from nicotinamide (NA) by the Preiss-Handler pathway [8]; importantly the principal pool of NAD+ in cells is recycled from nicotinamide (NAM) via the salvage pathway [7–9].

Paul Delgado-Olguín (ed.), *Mouse Embryogenesis*, Methods in Molecular Biology, vol. 3010,
https://doi.org/10.1007/978-1-0716-5126-1_17,

Interestingly, alterations in Nicotinamide phosphoribosyltransferase (NAMPT), the rate-limiting enzyme of salvage pathways or the *Nampt* gene, lead to a dramatic decrease in NAD+ bioavailability [10–12]. Meanwhile, the consumption of NAD+ by enzymes including PARPs, Sirtuins, CD38/CD157 and SARM1 does not significantly impact NAD+ intracellular concentrations, because NAM, the product of NAD+ consumers, is recycled via the salvage pathway [13, 14]. According with this mechanism, the experimental approach most commonly used to modify NAD+ bioavailability involves the specific inhibition of NAMPT activity using the specific inhibitor FK866 or by providing precursors, such as NA or NAM, to increase its intracellular concentrations [15, 16].

NAD^+ serves as an electron carrier in redox reactions, transitioning between its oxidized (NAD^+) and reduced (NADH) forms [17, 18]. It functions as an essential coenzyme for several critical proteins involved in glycolysis, the tricarboxylic acid cycle (TCA), and oxidative phosphorylation (OxPhos). As such, many energy metabolic pathways respond to changes in the redox state [19–22]. The redox state determines the oxidation state of an element, which influences its biological behavior and chemical reactivity in the intracellular environment [23]. Within the cell, the redox potential measures how oxidizing or reducing the cellular environment is, and this is largely influenced by the NAD^+/NADH ratio [24–26]. Hence, an increase in the bioavailability of NAD+ activates catabolic pathways and enhances the conversion of lactate to pyruvate. Conversely, NADH overproduction triggers anabolic pathways, while the electron transport chain (ETC) utilizes its redox potential to produce ATP [27, 28].

During early mouse development, each stage transition is accompanied by a reconfiguration of the cellular redox state, which is more reduced in the zygote and more oxidized in the blastocyst [29]. In this context, the reduced environment of the zygote promotes the generation of lactate coupling with NADH production, but in blastocyst the increased bioavailability of NAD+ enhances oxidative metabolic plasticity by coupling glycolysis and the Krebs cycle with OxPhos to meet the energy demands of active cell division [30].

The bioavailability of NAD+ also influences the epigenetic landscape through Sirtuins, a family of NAD+-dependent proteins that deacetylate histones [31]. Sirtuins reduce the acetylation of histones H4, H2B, and H3, particularly during zygotic genome activation (ZGA), a key developmental transition from maternal to embryonic control of development that occurs in minor and major phases [33]. Minor ZGA occurs in the early 2-cell stage to the 4-cell stage, where a subset of genes, often involved in basic cellular functions, are activated; and major ZGA begins later in development (around the 8-cell to blastocyst stages in mice), when a broader set of genes involved in development, cell differentiation, and more specialized functions are activated [32].

Additionally, other NAD+ consumers like Poly (ADP-ribose) polymerases (PARPs), link NAD+ with developmental processes increasing the number of protein isoforms during differentiation via control of the spliceosome complex by PARPs during major ZGA [34].

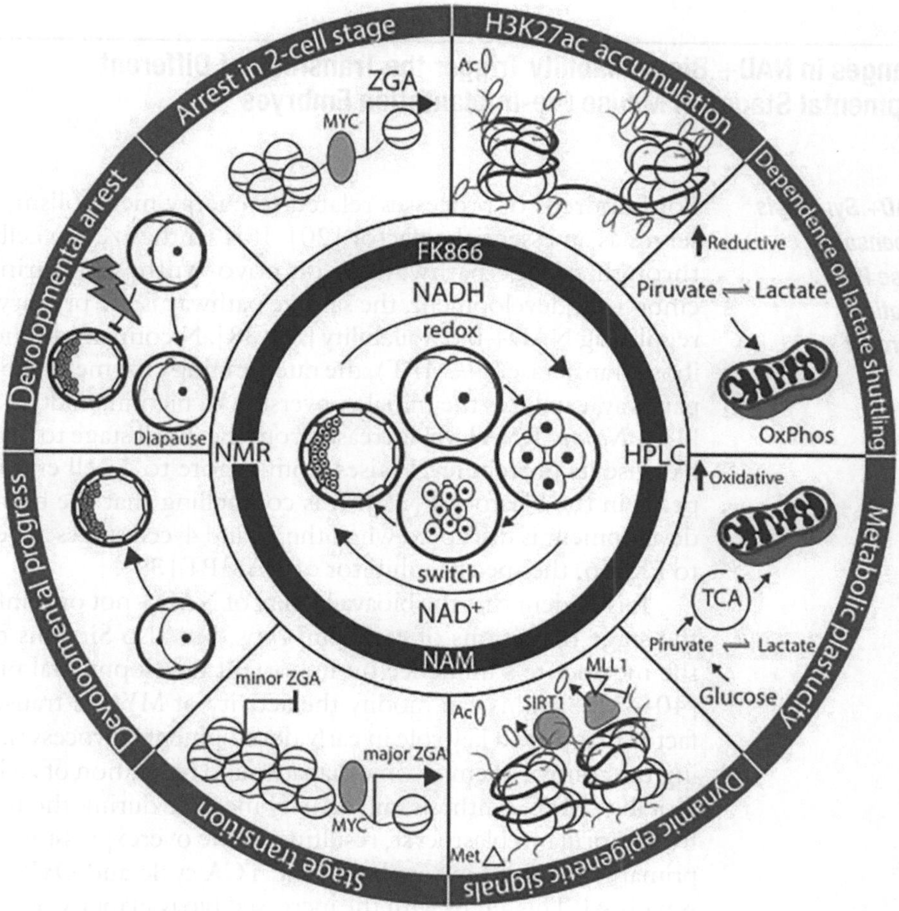

Fig. 1 The influence of NAD+ on development. The central scheme illustrates progression through the 2-, 4-, 8-cell, and blastocyst stages, accompanied by a redox shift in the NAD+/NADH cycle. Perturbation of NAD+ metabolism by FK866 leads to depletion and developmental arrest (upper half of the circle), whereas supplementation with nicotinamide (NAM) increases NAD+ availability and supports developmental progression (lower half). Surrounding panels highlight key NAD+-dependent processes: (i) Developmental arrest, with 2-cell stage block or diapause upon NAD+ depletion; (ii) zygotic genome activation (ZGA), where adequate NAD+ availability promotes transition from minor to major ZGA and supports MYC-dependent transcription; (iii) epigenetic regulation, with NAD+ availability influencing H3K27ac accumulation and histone methylation (MetΔ); (iv) metabolic plasticity, where NAD+ balance determines reductive pyruvate–lactate shuttling versus oxidative metabolism through OxPhos and the TCA cycle; and (v) epigenetic signaling, in which NAD+ fuels SIRT1- and MLL1-dependent chromatin modifications that coordinate transcriptional programs.

Since NAD+ plays a pivotal role as a regulator of key metabolic processes during early mouse development, disruptions in its synthesis could potentially induce embryonic diapause [35, 36]. Therefore, characterizing the dynamics of this metabolite could help to better understand the function of energy metabolism in embryonic development, and define the consequences of altering energy homeostasis. Moreover, this knowledge could inform strategies to manipulate energy metabolism and epigenetic systems to influence the early stages of development (Fig. 1).

2 Changes in NAD+ Bioavailability Trigger the Transition of Different Developmental Stages in Mouse Pre-Implantation Embryos

2.1 NAD+ Synthesis is Indispensable for Mouse Pre-Implantation Development

For hundreds of processes related to energy metabolism, NAD+ serves as an essential cofactor [20]. It is synthesized in cells either through salvage pathways or de novo synthesis. During early embryonic development, the salvage pathway is the primary source regulating NAD+ bioavailability [37, 38]. Nicotinamide phosphoribosyltransferase (NAMPT), the rate-limiting enzyme of the salvage pathway, catalyzes the initial conversion of nicotinamide to NAD+ [12]. *Nampt* RNA level increases from the 2-cell stage to blastocyst, likewise its protein level raises from zygote to 2-cell embryo and peaks in the blastocyst [39]. It is compelling that the blastocyst's development is disrupted when the 2- and 4-cell stages are exposed to FK866, the specific inhibitor of NAMPT [39].

It is evident that the bioavailability of NAD+ not only influences the stage transitions of early embryos, but also Sirtuins mediate the metabolite's influence on important developmental processes [40–42]. Sirtuins can modify the activity of MYC, a transcription factor that plays a key role in early developmental processes, including the control of embryonic diapause and regulation of ZGA [43]. Notably, MYC synthesis and activity increase during the transition from morula to blastocyst, resulting in the overexpression of genes primarily involved in regulating the TCA cycle and OxPhos pathways [44]. This aligns with the increased bioavailability of NAD+ in the blastocyst and the activation of SIRT1. When MYC is inhibited by NAD+ deprivation in the morula stage, blastocysts may enter a reversible latent state resembling embryonic diapause [39].

This suggests that the amount of NAD+ utilized during each specific embryonic phase in mice is tightly regulated, as suggested by the importance of NAD+ production in facilitating the transition from the 2-cell stage to the blastocyst. Consequently, a decrease in NAD+ bioavailability may significantly disrupt early developmental processes.

2.2 NAD+/NADH Ratio Defines Metabolic Transitions from Zygote to Blastocyst

Metabolism undergoes significant reprogramming during the transition from zygote to blastocyst [45–48]. The NAD+/NADH ratio is closely associated with these changes, as the bioavailability of the reduced (NADH) or oxidized (NAD+) forms determines which bioenergetic pathways are preferentially enhanced [49, 50].

Direct measurements of NAD+ and NADH using NMR spectroscopy or HPLC-based methodologies during different developmental phases could help examine their dynamics in the embryo [51–53]. In this context, NAD+ levels steadily increase during blastocyst development after a significant drop between the 1C (26 h) and 2C (48 h) stages [23]. Meanwhile, NADH levels progressively decrease throughout blastocyst development. Thus, the NAD+/NADH ratio declines in zygotes, steadily increases from the 2-cell stage to the morula, and reaches its highest levels in the blastocyst. Consequently, the overall NAD(H) pool is reduced in the early stages and becomes more oxidized as development progresses [31, 39].

The shift of energy sources in the mouse embryo is also influenced by the redox balance of the NAD+/NADH ratio [47, 48]. For example, reducing conditions in zygotes favor the switch from pyruvate to lactate as the primary energy source. However, later in the blastocyst stage, the embryo shifts to glucose metabolism [4, 5, 48]. Depending on the redox state, NAD+ consumption through glycolysis and the TCA cycle increases significantly during the transition from the 8-cell stage to blastocyst [54, 55]. This process couples oxidative pathways with OxPhos as the bioavailability of NAD+ rises.

Oxidative metabolism of carbohydrates is lower in early cleavage embryos compared to blastocysts and somatic cells [56, 57]. Glycolysis-related intermediates, including glucose, fructose-1,6-bisphosphate, and pyruvic acid, become more abundant in blastocysts, accompanied by a dramatic increase in the expression of glycolytic enzymes. These enzymes, such as *Hk, Gpi, Aldo, Gapdh, Pgk, Pgam, Pkm, Hk2, Tpi1, Eno1*, and the glucose transporter *Slc2a3*, are barely detectable in cleavage-stage embryos [31, 39, 58–60].

Notably, TCA cycle intermediate metabolites are present in significantly higher amounts in blastocyst embryos compared to those in the zygote and 2-cell stage, including citrate, succinate, and alpha-ketoglutarate (α-KG), alongside increased expression of TCA cycle genes such as *Aco1, Idh2, Sucla2, Sdha, Fh1, and Mdh2* [55, 61, 62]. The enhanced mitochondrial TCA cycle reflects the increased oxidative metabolism in blastocysts, aligning with the high energy demands required for active cell proliferation in the transition from zygote to morula [29].

The TCA cycle in blastocysts is further increased by the accumulation of substrates derived from various oxidative reactions [61]. Lactate dehydrogenase B (LDHB), the prevalent isoform in blastocysts, facilitates the synthesis of pyruvate from lactate in an oxidative environment [63]. Similarly, a high concentration of NAD+ and a low concentration of NADH favor the conversion of pyruvate to acetyl-CoA, as pyruvate dehydrogenase (PDH) utilizes NAD+ to oxidize pyruvate into acetyl-CoA [64].

Likewise, within the mitochondria, the conversion of malate to oxaloacetate sustains the TCA cycle flux [65, 66]. The enzyme malate dehydrogenase, which catalyzes the interconversion of malate and oxaloacetate, facilitates oxaloacetate synthesis when NAD+ levels increase in blastocysts [66].

According to alterations in the NAD+/NADH ratio, it is worth noting that pyruvate deficiency leads to an increase in cytoplasmic NADH, while lactate withdrawal raises NAD+ through the lactate dehydrogenase (LDH) enzyme reaction [67]. This mechanism is essential in maintaining the redox balance. Interestingly, embryos deprived of pyruvate at the zygote, 2-cell, or blastocyst stages cannot progress to the next stage, even when the culture media is supplemented with lactate [39]. However, when the redox state is restored by activating LDHC, which is a lactate dehydrogenase that metabolizes α-ketovaleric acid (a compound with no known function in mammalian cells but an effective substrate for LDHC), NADH is oxidized to NAD+. This restoration allows lactate to support development through a redox-mediated mechanism [39]. Lactate can replace pyruvate at any developmental stage, but only if the cytoplasmic redox balance is properly maintained, underscoring the central role of redox states during embryonic development.

The shift in cellular energy sources is influenced by the redox balance of the NAD+/NADH ratio [68]. Studies have shown that aspartate (Asp) is highly sensitive to the cell's redox state and is crucial for the malate-aspartate (Mal-Asp) shuttle, which mediates the interconversion of NAD+ and NADH [66]. Notably, Asp levels are 20 times higher in blastocysts than in 2-cell embryos. These significant changes in Asp levels during normal development may indicate a shift in energy sources driven by the NAD+/NADH ratio, through the coupling of NAD+-dependent enzyme reactions to pyruvate production.

The generation of a pool of aspartate as a backup for pyruvate is sustained by an increase in NAD+ levels in the blastocyst [69]. This oxidized environment promotes the rapid conversion of malate to oxaloacetate via malate dehydrogenase within the mitochondria. In the cytosol, high concentrations of aspartate are converted to oxaloacetate by aspartate aminotransferase (AST). Next, oxaloacetate is converted to malate, which is exported to the mitochondria and reconverted to oxaloacetate to drive the TCA cycle [70]. Additionally, gene expression analysis reveals that the enzymes involved in the malate-aspartate (Mal-Asp) shuttle are expressed dynamically during normal development. Between the 2-cell and blastocyst stages, the cytoplasmic enzyme Malate dehydrogenase 1 (MDH1) and the mitochondrial enzyme Malate dehydrogenase 2 (MDH2) increase by approximately 20-fold and 4-fold, respectively, while the mitochondrial enzymes AST1 and AST2 increase by 9-fold and 2.5-fold, respectively [47]. This indicates that the embryo's metabolic plasticity enables redox control mechanisms to shift from LDH activity in the zygote to Mal-Asp shuttle activity in the blastocyst [65].

The development of the zygote, 2-cell, and morula depends on maintaining a redox state balance [71, 72], as both glycolysis and the TCA cycle rely on NAD+ reduction to sustain the energy supply. However, metabolic plasticity increases in the embryo as it develops, enabling utilization of a broader range of pathways for energy production [73] (Fig. 1).

In contrast to zygote and 2-cell stages, fatty acid metabolism increases significantly in blastocysts, allowing fatty acids alone to sustain blastocyst development [74]. The blastocyst's increased tolerance to alterations in the redox state is associated with its enhanced metabolic adaptability. This is supported by the expression of acyl-CoA dehydrogenase genes (*Acadm*, *Acadl*, and *Acadvl*), which catalyze the initial step in mitochondrial β-oxidation. These genes are expressed at much higher levels in late morulae and blastocysts compared to early cleavage stages, resulting in a substantial increase in mitochondrial fatty acid β-oxidation during preimplantation development [75, 76]. The elevated expression of fatty acid oxidation (FAO) enzymes during development highlights the role of an oxidized environment in metabolic reprogramming, enabling the use of a broader range of energy resources as the embryo matures.

2.3 NAD+ Metabolism and its Impact on Epigenetic Remodeling

Recently, the role of NAD+ in shaping the epigenetic landscape has gained significant attention, particularly through its connection with sirtuins, a group of NAD+-dependent deacetylases [77–79]. Sirtuins directly influence the acetylation of lysine 27 of histone H3(H3K27) [80, 81] and affect the methylation of histone H3 at K4 by regulating the activity of MLL1, a histone methyltransferase (HMT) involved in transcriptional activation [82].

Through chromatin immunoprecipitation sequencing (ChIP-seq) and immunostaining of H3K27ac across the zygote, 2-cell, 4-cell, morula, and blastocyst stages, the distribution of H3K9ac did not exhibit any notable changes [83]. However, H3K27ac underwent significant modifications during mouse preimplantation development [84, 85]. The formation of the zygote is marked by a rapid increase in H3K27ac levels in the intergenic, intronic, and promoter regions of genes associated with zygotic stage maintenance [86, 87]. In contrast, H3K27ac levels decreased significantly during the late 2-cell and 4-cell stages, continued to decline in the morula, and were ultimately lost in the blastocyst stage [86]. These dynamic changes align with shifts in developmental transcriptional programs. Interestingly, NAD+ depletion using FK866 resulted in sustained high levels of

H3K27ac at the late 2-cell and 4-cell stages, which was reversed by supplementation with NMN. This suggests that NAD+ production plays a critical role in the deacetylation of H3K27ac during cleavage development in the mouse embryo.

Genes in the embryo are expressed for the first time after fertilization in a process known as zygotic genome activation (ZGA). ZGA is a critical event in development that marks the transition from maternal to embryonic control of gene expression [88]. ZGA usually entails two waves of transcription, minor ZGA, when a small set of genes are activated, and major ZGA, when thousands of genes are transcribed [89].

In mouse, minor ZGA and major ZGA occur during middle-1-cell and late 2-cell stages, respectively. NAD+ plays a critical role in zygotic genome activation (ZGA) by cooperating with SIRT1 to remove the H3K27ac mark from promoters of minor ZGA genes [90], facilitating the transition from higher levels of this mark in the early 2-cell embryo to significantly lower levels in the late 2-cell embryo [32]. At the mid-2-cell stage, SIRT1 is highly expressed and facilitates the ZGA through NAD+-mediated deacetylation. Notably, the ability to progress to the blastocyst stage is significantly impaired in SIRT1-depleted embryos, with development primarily arresting at the morula stage. Furthermore, at the late 2-cell stage, embryos lacking SIRT1 exhibit persistently elevated levels of H3K27ac, highlighting SIRT1's essential role in regulating these epigenetic changes [32].

DNA demethylation preserves naïve pluripotency and promotes blastocyst development [91]. TET enzymes, which catalyze DNA demethylation by successive oxidation of 5mC, rely on the metabolic intermediate α-KG as a cofactor [92]. An increase in α-KG through oxidative pathways promotes an increase in the number of embryoblasts cells and supports blastocyst formation in vitro by activating DNA demethylation TET enzymes [93]. Increased α-KG is accompanied by an increase in NAD+ levels during the blastocyst stage, as NAD+ facilitates the conversion of D-isocitrate to α-KG via isocitrate dehydrogenase [3]. Moreover, SIRT1 deacetylates TET2, enhancing its activity and coupling the increased bioavailability of TCA cycle substrates and NAD+ with DNA demethylation during the blastocyst stage [94] (Fig. 1).

L-2-hydroxyglutarate (L-2-HG), a metabolite recently identified in early embryos [95], functions as an antagonist of α-KG, preventing histone demethylation [96, 97]. As previously mentioned, α-KG is essential for the enzymatic activity of numerous dioxygenases, including histone demethylases. MII oocytes, zygotes, and 2-cell embryos exhibit the highest levels of L-2-HG, which gradually decline as the embryo develops, while α-KG levels increase, reaching their peak in the blastocyst stage [47]. Induced elevation of L-2-HG during the zygote and 2-cell stages resulted in developmental abnormalities and reduced blastocyst formation, accompanied by decreased expression of pluripotency and TCA cycle genes. Conversely, treatment with embryo-permeable dimethyl-α-KG at tolerable concentrations induces an increased expression of these genes [47].

These findings suggest that α-KG production via isocitrate dehydrogenase, coupled with NAD+ reduction, is sustained by bioavailable NAD+. Additionally, α-KG is produced through the malate-alanine shuttle, which supports blastocyst development and facilitates global histone methylation removal [66]. During the zygote and 2-cell stages, low NAD+ levels correlate with reduced α-KG and elevated L-2-HG levels, thereby preventing epigenetic remodeling in zygote until the activation of ZGA [47]. This alteration in metabolite balance impedes the transition required for ZGA activation. Once NAD+ levels increase, this shift facilitates the activation of the necessary enzymes, like SIRT1 and TET, promoting the correct epigenetic changes in the next stage.

The shift from high L-2-HG to high α-KG levels aligns with a transition from a reduced to an oxidative state [70], as histone marks removal during early embryogenesis requires this reduction in L-2-HG. These findings underscore the influence of the redox state on metabolic control and epigenome remodeling during embryonic development.

2.4 Deprivation of NAD+ Impairs Blastocyst Development

Early stages of development are governed by metabolic dynamics, with particular emphasis on the role of NAD+ bioavailability [98]. Impaired NAD+ production in early embryos leads to developmental arrest, preventing progression to the next stage [99]. Notably, treatment with nicotinamide mononucleotide (NMN) completely restored development in embryos treated with FK866, which otherwise exhibited total arrest between the 4- and 8-cell stages. Also, blocking NAD+ production with FK866 significantly reduced early cell division and hindered blastocyst development [32].

The direction of the reaction that interconverts pyruvic acid and lactate via LDHB contributes to maintaining the balance between free NAD+ and NADH in the cytosol. Disruption of this mechanism negatively impacts embryonic development. Zygotes treated with oxamate, a specific LDHB inhibitor, exhibited developmental arrest at the 8- to 16-cell stages. However, this block was successfully reversed by the addition of NMN, which restored the redox balance [47].

Additionally, the FK866-treated late 2-cell embryo exhibited a substantial upregulation of Duxf3 (Dux), a master transcription factor that uniquely expresses itself in early mouse 2-cell embryo and transiently regulates the ZGA process. Genes that are important for setting up early embryonic identity and ensuring proper progression through early cell divisions, considered part of the minor ZGA gene pool, exhibit extended expression until the late 2- and 4-cell stages, a process mediated by the reduced activity of SIRT1 when NAD+ bioavailability is deficient [47].

Interestingly, treatment with FK866 maintained the expression of minor ZGA genes beyond the 4-cell stage, highlighting the critical role of NAD+ synthesis in regulating the timely transition between the transcriptional programs of minor ZGA to major ZGA. Supplementation with NMN at the 4-cell stage restored the proper transition between minor and major ZGA genes. Aberrant minor ZGA, such as the prolonged activation of Dux6 at the 4-cell stage, severely disrupts major ZGA and embryonic development [47]. These findings suggest that excessive activation of minor ZGA contributes to developmental failure following NAD+ depletion (Fig. 1).

Overall, the evidence underscores the requirement of bioavailable NAD+ for the successful progression of early embryonic development. Therefore, elucidating how NAD+ concentration dynamics is regulated in early embryos is fundamental for understanding the metabolic control of embryogenesis.

Acknowledgments

This work was supported by the Canadian Institutes of Health Research (CIHR) (162208, 149046 and 468633) and the Heart & Stroke Foundation of Canada (G-25-0040100) to P.D.-O.; and by the SickKids Research Institute Strategic Fund, Restracomp (SickKids), the Labatt Family Heart Centre (SickKids), and the Ted Rogers Centre for Heart Research, to Q.E.-C.

Conflict of Interest The authors have no conflicts of interest to declare that are relevant to the content of this chapter.

References

1. Rivera-Pérez JA, Hadjantonakis AK (2014) The dynamics of morphogenesis in the early mouse embryo. Cold Spring Harb Perspect Biol 7(11):a015867
2. Kojima Y, Tam OH, Tam PP (2014) Timing of developmental events in the early mouse embryo. Semin Cell Dev Biol 34:65–75
3. Zhao J, Yao K, Yu H, Zhang L, Xu Y, Chen L, Sun Z, Zhu Y, Zhang C, Qian Y, Ji S, Pan H, Zhang M, Chen J, Correia C, Weiskittel T, Lin DW, Zhao Y, Chandrasekaran S, Fu X, Zhang D, Fan HY, Xie W, Li H, Hu Z, Zhang J (2021) Metabolic remodelling during early mouse embryo development. Nat Metab 3(10):1372–1384
4. Chason RJ, Csokmay J, Segars JH, DeCherney AH, Armant DR (2011) Environmental and epigenetic effects upon preimplantation embryo metabolism and development. Trends Endocrinol Metab 22(10):412–420
5. Xu J, Sinclair KD (2015) One-carbon metabolism and epigenetic regulation of embryo development. Reprod Fertil Dev 27(4):667–676
6. Xiao W, Wang RS, Handy DE, Loscalzo J (2018) NAD (H) and NADP (H) redox couples and cellular energy metabolism. Antioxid Redox Signal 28(3):251–272
7. Imai S-Ichiro (2009) The NAD World: a new systemic regulatory network for metabolism and aging—Sirt1, systemic NAD biosynthesis, and their importance. Cell Biochem Biophys 53:65–74
8. Imai S-Ichiro (2016) The NAD World 2.0: the importance of the inter-tissue communication

mediated by NAMPT/NAD+/SIRT1 in mammalian aging and longevity control. NPJ Syst Biol Appl 2(1):1–9
9. Mouchiroud L, Houtkooper RH, Auwerx J (2013) NAD+ metabolism: a therapeutic target for age-related metabolic disease. Crit Rev Biochem Mol Biol 48(4):397–408
10. Burgos E S (2011) NAMPT in regulated NAD biosynthesis and its pivotal role in human metabolism. Curr Med Chem 18(13):1947–1961
11. Garten A, Schuster S, Penke M, Gorski T, De Giorgis T, Kiess W (2015) Physiological and pathophysiological roles of NAMPT and NAD metabolism. Nat Rev Endocrinol 11(9):535–546
12. Imai S-I (2009) Nicotinamide phosphoribosyltransferase (Nampt): a link between NAD biology, metabolism, and diseases. Curr Pharm Des 15(1):20–28
13. Ichiro I-S (2011) Dissecting systemic control of metabolism and aging in the NAD World: the importance of SIRT1 and NAMPT-mediated NAD biosynthesis. FEBS Lett 585(11):1657–1662
14. Hayashida S, Arimoto A, Kuramoto Y, Kozako T, Honda S-Ichiro, Shimeno H et al (2010) Fasting promotes the expression of SIRT1, an NAD+-dependent protein deacetylase, via activation of PPARα in mice. Mol Cell Biochem 339:285–292
15. Song TY, Yeh SL, Hu ML, Chen MY, Yang NC (2015) A Nampt inhibitor FK866 mimics vitamin B3 deficiency by causing senescence of human fibroblastic Hs68 cells via attenuation of NAD+-SIRT1 signaling. Biogerontology 16:789–800
16. Ferguson N (2017) Characterisation of pharmacological mechanisms and potential therapeutic uses of FK866. 2017 [cited 2025 Jan 22]. Available from: https://plymouth.researchcommons.org/foh-theses-other/199/
17. Di Stefano M, Conforti L (2013) Diversification of NAD biological role: the importance of location. FEBS J 280(19):4711–4728
18. Opitz CA, Heiland I (2015) Dynamics of NAD-metabolism: everything but constant. Biochem Soc Trans 43(6):1127–1132
19. Strømland Ø, Niere M, Nikiforov AA, VanLinden MR, Heiland I, Ziegler M (2019) Keeping the balance in NAD metabolism. Biochem Soc Trans 47(1):119–130
20. Stein LR, Imai S-I (2012) The dynamic regulation of NAD metabolism in mitochondria. Trends Endocrinol Metab 23(9):420–428.
21. Houtkooper RH, Cantó C, Wanders RJ, Auwerx J (2010) The secret life of NAD+: an old metabolite controlling new metabolic signaling pathways. Endocr Rev 31(2):194–223
22. Garten A, Schuster S, Penke M, Gorski T, De Giorgis T, Kiess W (2015) Physiological and pathophysiological roles of NAMPT and NAD metabolism. Nat Rev Endocrinol 11(9):535–546
23. Dumollard R, Ward Z, Carroll J, Duchen MR (2007) Regulation of redox metabolism in the mouse oocyte and embryo. Development (2007) 134 (3): 455–465
24. Vina J, Saez GT, Gambini J, Gomez-Cabrera MC, Borrás C (2016) Role of NAD (+)/NADH redox ratio in cell metabolism: a tribute to Helmut Sies and Theodor Bücher and Hans A. Krebs. Arch Biochem Biophys 595:176–180
25. Xiao W, Wang RS, Handy DE, Loscalzo J (2018) NAD (H) and NADP (H) redox couples and cellular energy metabolism. Antioxid Redox Signal 28(3):251–272
26. Blacker TS, Duchen MR (2016) Investigating mitochondrial redox state using NADH and NADPH autofluorescence. Free Radic Biol Med 100:53–65
27. Shin IH, Jeon SJ, Park HS, Park DH (2004) Catalytic oxidoreduction of pyruvate/lactate and acetaldehyde/ethanol coupled to electrochemical oxidoreduction of NAD+ /NADH. J Microbiol Biotechnol 14(3):540–546
28. Dawson AG (1979) Oxidation of cytosolic NADH formed during aerobic metabolism in mammalian cells. Trends Biochem Sci 4(8):171–176
29. Harvey AJ, Kind KL, Thompson JG (2002) REDOX regulation of early embryo development. Reproduction 123(4):479–486
30. Bedzhov I, Graham SJ, Leung CY, Zernicka-Goetz M (2014) Developmental plasticity, cell fate specification and morphogenesis in the early mouse embryo. Philos Trans R Soc B Biol Sci 369(1657):20130538
31. Verdin E (2014) The many faces of sirtuins: coupling of NAD metabolism, sirtuins and lifespan. Nat Med 20(1):25–27
32. Li J, Zhang J, Hou W, Yang X, Liu X, Zhang Y et al (2022) Metabolic control of histone acetylation for precise and timely regulation of minor ZGA in early mammalian embryos. Cell Discov 8(1):96
33. Wu J, Xu J, Liu B, Yao G, Wang P, Lin Z et al (2018) Chromatin analysis in human early development reveals epigenetic transition during ZGA. Nature 557(7704):256–260
34. Jones A, Kraus WL (2022) Multiomics analysis of the NAD+–PARP1 axis reveals a role for site-specific ADP-ribosylation in splicing in embryonic stem cells. Genes Dev 36(9–10):601–617
35. Tippetts TS, Sieber MH, Solmonson A (2023) Beyond energy and growth: the role of metabolism in developmental signaling, cell behavior and diapause. Development 150(20):dev201610
36. Zhao J, Wang W, Zhang L, Zhang J, Sturmey R, Zhang J (2023) Dynamic metabolism during early mammalian embryogenesis. Development 150(20):dev202148

37. Cuny H, Rapadas M, Gereis J, Martin EM, Kirk RB, Shi H et al (2020) NAD deficiency due to environmental factors or gene–environment interactions causes congenital malformations and miscarriage in mice. Proc Natl Acad Sci 117(7):3738–3747
38. Warren A, Porter RM, Reyes-Castro O, Ali MM, Marques-Carvalho A, Kim HN et al (2023) The NAD salvage pathway in mesenchymal cells is indispensable for skeletal development in mice. Nat Commun 14(1):3616
39. Sharpley MS, Chi F, Ten Hoeve J, Banerjee U (2021) Metabolic plasticity drives development during mammalian embryogenesis. Dev Cell 56(16):2329–2347
40. Nevoral J, Drutovic D, Vaskovicova M, Benc M, Liska F, Valentova I et al (2024) Dynamics and necessity of SIRT1 for maternal–zygotic transition. Sci Rep 14(1):21598.
41. Nevoral J, Sutovsky P, Zamostna K, Petelak A, Hoskova K, Zalmanova T et al (2015) SIRT1 as a key factor for histone code establishment in early embryo, from a perspective of assisted reproduction. Slovak J Anim Sci 48(4):145–158
42. Hwang YJ, Sung GJ, Marquardt R, Young SL, Lessey BA, Kim TH et al (2022) SIRT1 plays an important role in implantation and decidualization during mouse early pregnancy. Biol Reprod 106(6):1072–1082
43. Fan W, Li X (2023) The SIRT1-c-Myc axis in regulation of stem cells. Front Cell Dev Biol 11:1236968
44. Tang S, Fang Y, Huang G, Xu X, Padilla-Banks E, Fan W et al (2017) Methionine metabolism is essential for SIRT 1-regulated mouse embryonic stem cell maintenance and embryonic development. EMBO J 36(21):3175–3193
45. Zhao J, Yao K, Yu H, Zhang L, Xu Y, Hu Z et al (2021) Metabolic remodeling during murine early embryo development. Nat Metab. (10):1372-1384
46. Xu J, Sinclair KD (2015) One-carbon metabolism and epigenetic regulation of embryo development. Reprod Fertil Dev 27(4):667–676
47. Miyazawa H, Aulehla A (2018) Revisiting the role of metabolism during development. Development 145(19):dev131110
48. Chason RJ, Csokmay J, Segars JH, DeCherney AH, Armant DR (2011) Environmental and epigenetic effects upon preimplantation embryo metabolism and development. Trends Endocrinol Metab 22(10):412–420
49. Titov DV, Cracan V, Goodman RP, Peng J, Grabarek Z, Mootha VK (2016) Complementation of mitochondrial electron transport chain by manipulation of the NAD+ /NADH ratio. Science 352(6282):231–235
50. Anderson KA, Madsen AS, Olsen CA, Hirschey MD (2017) Metabolic control by sirtuins and other enzymes that sense NAD+ , NADH, or their ratio. Biochim Biophys Acta 1858(12):991–998
51. Yamada K, Hara N, Shibata T, Osago H, Tsuchiya M (2006) The simultaneous measurement of nicotinamide adenine dinucleotide and related compounds by liquid chromatography/ electrospray ionization tandem mass spectrometry. Anal Biochem 352(2):282–285
52. Yaku K, Okabe K, Nakagawa T (2018) Simultaneous measurement of NAD metabolome in aged mice tissue using liquid chromatography tandem-mass spectrometry. Biomed Chromatogr 32(6):e4205
53. Kern SE, Price-Whelan A, Newman DK (2014) Extraction and measurement of NAD (P)+ and NAD (P) H. Methods Mol Biol 2014(1149):311–323
54. Stern S, Biggers JD, Anderson E (1971) Mitochondria and early development of the mouse. J Exp Zool 176(2):179–191
55. Nagaraj R, Sharpley MS, Chi F, Braas D, Zhou Y, Kim R et al (2017) Nuclear localization of mitochondrial TCA cycle enzymes as a critical step in mammalian zygotic genome activation. Cell 168(1):210–223
56. Wells PG, Miller-Pinsler L, Shapiro AM (2014) Impact of oxidative stress on development. In: Perinatal and prenatal disorders. Springer New York, New York, NY, pp 1–37
57. Dennery PA (2007) Effects of oxidative stress on embryonic development. Birth Defects Res Part C Embryo Today Rev 81(3):155–162
58. Sharov AA, Piao Y, Matoba R, Dudekula DB, Qian Y, VanBuren V et al (2003) Transcriptome analysis of mouse stem cells and early embryos. PLoS Biol 1(3):e74
59. Park SJ, Shirahige K, Ohsugi M, Nakai K (2015) DBTMEE: a database of transcriptome in mouse early embryos. Nucleic Acids Res 43(D1):D771–D776
60. Diez-Roux G, Banfi S, Sultan M, Geffers L, Anand S, Rozado D et al (2011) A high-resolution anatomical atlas of the transcriptome in the mouse embryo. PLoS Biol 9(1):e1000582
61. Hong J, Tong H, Wang X, Lv X, He L, Yang X et al (2023) Embryonic diapause due to high glucose is related to changes in glycolysis and oxidative phosphorylation, as well as abnormalities in the TCA cycle and amino acid metabolism. Front Endocrinol 14:1135837
62. Heinemann-Yerushalmi L, Bentovim L, Felsenthal N, Vinestock RC, Michaeli N, Krief S et al (2021) BCKDK regulates the TCA cycle through PDC in the absence of PDK family during embryonic development. Dev Cell 56(8):1182–1194
63. Lee SH, Liu X, Jimenez-Morales D, Rinaudo PF (2022) Murine blastocysts generated by

in vitro fertilization show increased Warburg metabolism and altered lactate production. Elife 11:e79153
64. Gardner DK, Harvey AJ (2015) Blastocyst metabolism. Reprod Fertil Dev 27(4):638–654
65. Yoon SJ, Koo DB, Park JS, Choi KH, Han YM, Lee KA (2006) Role of cytosolic malate dehydrogenase in oocyte maturation and embryo development. Fertil Steril 86(4):1129–1136
66. Lane M, Gardner DK (2005) Mitochondrial malate-aspartate shuttle regulates mouse embryo nutrient consumption. J Biol Chem 280(18):18361–18367
67. Monk M, Ansell J (1976) Patterns of lactic dehydrogenase isozymes in mouse embryos over the implantation period in vivo and in vitro. Development 36(3):653–662
68. Ufer C, Wang CC, Borchert A, Heydeck D, Kuhn H (2010) Redox control in mammalian embryo development. Antioxid Redox Signal 13(6):833–875
69. Borst P (2020) The malate–aspartate shuttle (Borst cycle): How it started and developed into a major metabolic pathway. IUBMB Life 72(11):2241–2259
70. Mitchell M, Cashman KS, Gardner DK, Thompson JG, Lane M (2009) Disruption of mitochondrial malate-aspartate shuttle activity in mouse blastocysts impairs viability and fetal growth. Biol Reprod 80(2):295–301
71. Hardy ML, Day ML, Morris MB (2021) Redox regulation and oxidative stress in mammalian oocytes and embryos developed in vivo and in vitro. Int J Env Res Public Health 18(21):11374
72. Dumollard R, Carroll J, Duchen MR, Campbell K, Swann K (2009) Mitochondrial function and redox state in mammalian embryos. Semin. Cell Dev. Biol. (Vol 20, No 3), pp 346–353)
73. Krisher RL, Prather RS (2012) A role for the Warburg effect in preimplantation embryo development: metabolic modification to support rapid cell proliferation. Mol Reprod Dev 79(5):311–320
74. Dunning KR, Cashman K, Russell DL, Thompson JG, Norman RJ, Robker RL (2010) Beta-oxidation is essential for mouse oocyte developmental competence and early embryo development. Biol Reprod 83(6):909–918
75. Sturmey RG, Reis A, Leese HJ, McEvoy TG (2009) Role of fatty acids in energy provision during oocyte maturation and early embryo development. Reprod Domest Anim 44:50–58
76. McKeegan PJ, Sturmey RG (2011) The role of fatty acids in oocyte and early embryo development. Reprod Fertil Dev 24(1):59–67
77. Kosciuk T, Wang M, Hong JY, Lin H (2019) Updates on the epigenetic roles of sirtuins. Curr Opin Chem Biol 51:18–29
78. Jing H, Lin H (2015) Sirtuins in epigenetic regulation. Chem Rev 115(6):2350–2375
79. Bosch-Presegué L, Vaquero A (2015) Sirtuin-dependent epigenetic regulation in the maintenance of genome integrity. FEBS J 282(9):1745–1767
80. Wu Y, Liu Y, Hao Z, Liu X (2022) NAD+ is critical for maintaining acetyl-CoA and H3K27ac in embryonic stem cells by Sirt1-dependent deacetylation of AceCS1. Life Med 1(3):401–405
81. Hsu WW, Wu B, Liu WR (2016) Sirtuins 1 and 2 are universal histone deacetylases. ACS Chem Biol 11(3):792–799
82. Aguilar-Arnal L, Katada S, Orozco-Solis R, Sassone-Corsi P (2015) NAD+-SIRT1 control of H3K4 trimethylation through circadian deacetylation of MLL1. Nat Struct Mol Biol 22(4):312–318
83. Bošković A, Bender A, Gall L, Ziegler-Birling C, Beaujean N, Torres-Padilla ME (2012) Analysis of active chromatin modifications in early mammalian embryos reveals uncoupling of H2A. Z acetylation and H3K36 trimethylation from embryonic genome activation. Epigenetics 7(7):747–757
84. Wang M, Chen Z, Zhang Y (2022) CBP/p300 and HDAC activities regulate H3K27 acetylation dynamics and zygotic genome activation in mouse preimplantation embryos. EMBO J 41(22):e112012
85. Chen Z, Djekidel MN, Zhang Y (2021) Distinct dynamics and functions of H2AK119ub1 and H3K27me3 in mouse preimplantation embryos. Nat Genet 53(4):551–563
86. Zhang T, Zhang Z, Dong Q, Xiong J, Zhu B (2020) Histone H3K27 acetylation is dispensable for enhancer activity in mouse embryonic stem cells. Genome Biol 21:1–7
87. Fu X, Zhang C, Zhang Y (2020) Epigenetic regulation of mouse preimplantation embryo development. Curr Opin Genet Dev 64:13–20
88. Yao C, Zhang W, Shuai L (2019) The first cell fate decision in pre-implantation mouse embryos. Cell Regen 8(2):51–57
89. Chen Y, Wang L, Guo F, Dai X, Zhang X (2023) Epigenetic reprogramming during the maternal-to-zygotic transition. MedComm 4(4):e331
90. Kafri T, Gao X, Razin A (1993) Mechanistic aspects of genome-wide demethylation in the preimplantation mouse embryo. Proc Natl Acad Sci USA 90(22):10558–10562
91. Zhang Z, He C, Zhang L, Zhu T, Lv D, Li G et al (2019) Alpha-ketoglutarate affects murine embryo development through metabolic and epigenetic modulations. Reproduction 158(2):123–133

92. Liu J, Chu M, Zhang J, He J, Yang Q, Tao L et al (2024) Glutathione safeguards TET-dependent DNA demethylation and is critical for the acquisition of totipotency and pluripotency during preimplantation development. FASEB J 38(3):e23453
93. Rzem R, Achouri Y, Marbaix E, Schakman O, Wiame E, Marie S et al. (2015) A mouse model of L-2-hydroxyglutaric aciduria, a disorder of metabolite repair. PLoS ONE 10(3):e0119540
94. Sun J, He X, Zhu Y, Ding Z, Dong H, Feng Y, Li L (2018) SIRT1 activation disrupts maintenance of myelodysplastic syndrome stem and progenitor cells by restoring TET2 function. Cell Stem Cell 23(3):355–369
95. Ye D, Guan KL, Xiong Y (2018) Metabolism, activity, and targeting of D-and L-2-hydroxyglutarates. Trends Cancer 4(2):151–165
96. Du X, Hu H (2021) The roles of 2-hydroxyglutarate. Front Cell Dev Biol 9:651317
97. Harvey AJ, Kind KL, Thompson JG (2002) REDOX regulation of early embryo development. Reproduction 123(4):479–486
98. Tsai FC, Gardner DK (1994) Nicotinamide, a component of complex culture media, inhibits mouse embryo development in vitro and reduces subsequent developmental potential after transfer. Fertil Steril 61(2):376–382
99. Newman H, Mercer S, Catt S, Horta F (2023) Assessing mouse embryo metabolic activity and metabolic modulation through NAD(P)H with non-invasive adapted confocal microscopy. Fertil Steril 120(4):e300

Chapter 18

3D Chromatin Architecture During Early Development: New Methods and New Findings

Eduardo Blanco-Olais, Karina Jácome-López, Rosario Pérez-Molina and Mayra Furlan-Magaril

Abstract

Studies on genome spatial organization have increased considerably in the past few decades, allowing for the detailed characterization of diverse genomic topological landscapes. Chromatin interactions between promoters and regulatory elements are crucial in controlling gene regulation during development, resulting in specific gene expression profiles characterized by transcript dosage and their distribution in time and space. Here, we describe both proximity-ligation-based and non-proximity-ligation-based new technologies, as well as their exciting findings in pluripotent cells and early embryogenesis. We also discuss examples of chromatin conformation disruption in disease and cancer.

Keywords Chromatin architecture, Genome topology, Distal regulatory elements

1 Introduction

Genome organization has been extensively studied, emphasizing the growing importance of genome structure at different scales. This growing interest is driven by the improvement of new high-resolution techniques that have uncovered new chromatin structures and revealed the molecular basis of genome topology. These findings result from improvements in ligation-dependent protocols, as well as the development of ligation-independent techniques that further enhance our understanding of genome topology and the significance of chromatin structure. Additionally, technologies that profile RNA–DNA interactions and FISH-based imaging techniques have been crucial to further measure topology dynamics. In this review, we summarize the most recent methods available to study genome 3D conformation and discuss recent findings in pluripotent cells, early embryogenesis, developmental disorders, and some mechanisms of topology disruption in cancer.

Paul Delgado-Olguín (ed.), *Mouse Embryogenesis*, Methods in Molecular Biology, vol. 3010,
https://doi.org/10.1007/978-1-0716-5126-1_18,

2 Technologies to Study Chromatin Architecture

2.1 Proximity Ligation-Based Technologies

Genome topology has been widely studied initially using microscopy to resolve chromosome territories and locate specific genomic loci [1, 2]. Subsequently, the introduction of the chromosome conformation capture (3C) assay by Dekker et al. [3] uncovered chromatin interactions in 3D at a higher resolution, and soon after, a variety of techniques based on the 3C method, including 4C [4, 5], 5C [6], and Hi-C [7], revealed chromatin structures at different scales. From these methods, approaches within the 3C family have been developed to understand chromatin structure at unprecedented resolutions.

These new techniques mainly represent improvements in the fixation, digestion, and data analysis (Fig. 1). For example, Micro-C uses a micrococcal Mnase and DGS/EGS additional crosslinkers instead of restriction enzymes (MboI, Dpnll) and only formaldehyde to obtain genome-wide nucleosome-resolution chromatin contacts [8]. Micro-C was first applied in yeast and then in mammalian cells, in which it has been used to characterize chromatin loops in detail [8, 9]. Another technique is Hi-C 3.0, which combines different cross-linking agents with multiple restriction enzymes, detecting more structures than conventional Hi-C [7]. In fact, a systematic evaluation by Oksuz et al. suggests that, depending on the study, you may choose your protocol, for example, Micro-C for loop detection at nucleosome resolution and microcompartments, or Hi-C 3.0 for detection of both loops (less resolution than MicroC but more than Hi-C) and larger compartments [10].

Also, other variants include computational techniques for analysis, such as HiCO 3D (high-throughput chromosome conformation capture with nucleosome orientation), which uses an MNase for digestion and with bioinformatic analysis, depending on the mapping direction of the reads, 3D positions, and orientations of every nucleosome can be obtained genome-wide [11].

Another exciting variant of 3C methods is called C-walks (Chromosomal walks). This technique involves selecting high-molecular-weight 3C DNA, performing serial dilutions, and distributing the DNA into wells. Following isothermal amplification and other steps, including sequencing and computational assembly, high-order structures are linked to multiple genomic loci to form proximity chains. This process reveals multi-way chromatin interactions [12].

In turn, most of these interesting techniques already mentioned have been modified for specific regions in a non-genome-wide version to increase their resolution. The Capture-C family of experiments uses oligonucleotide probes to enrich for specific regions [13], as

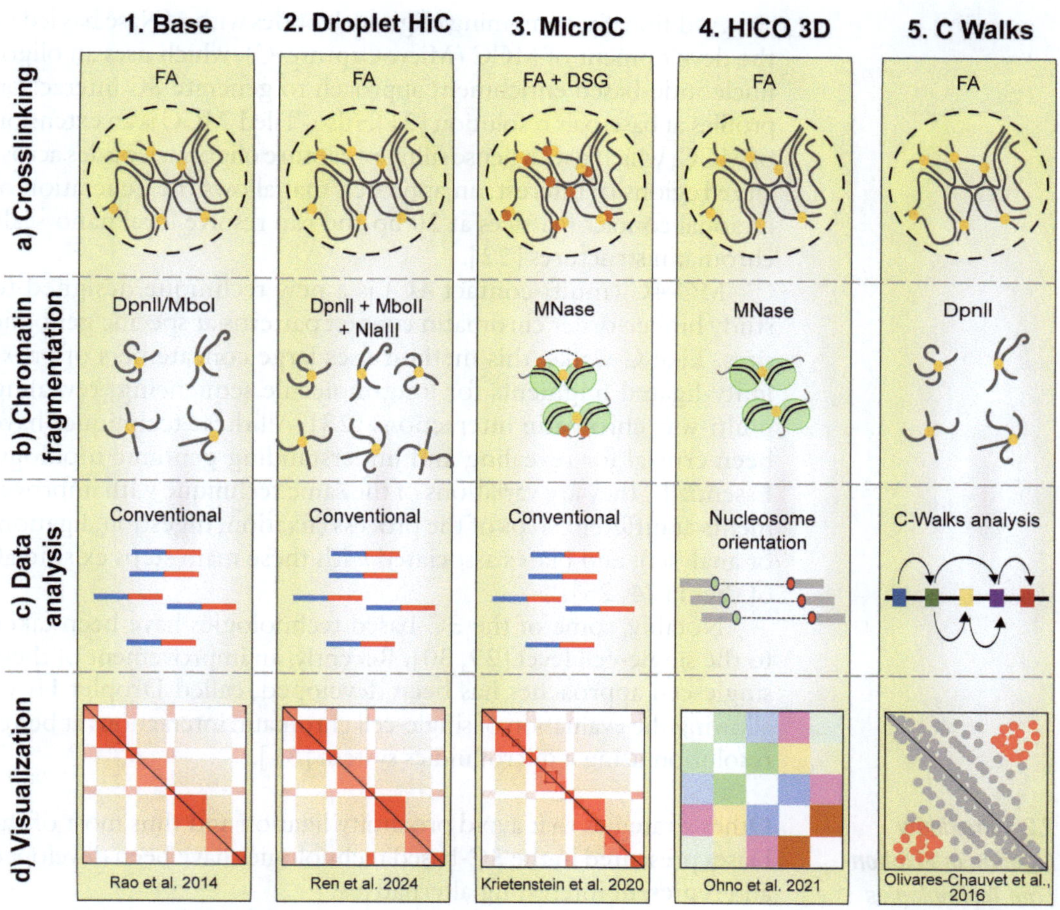

Fig. 1 Improvements in 3C-based technologies. The vertical axis (left) shows the main steps of the protocols, and in the figure, the modifications are indicated: (**a**) Crosslinking, (**b**) Chromatin fragmentation, (**c**) Data analysis, (**d**) Visualization. The name of the technology is at the top of each panel, and the modifications are indicated in each step. (1) Base 3C protocol in which the crosslink is done with formaldehyde (FA) represented by yellow dots, the digestion with a four-six cutter enzyme (MboI or DpnII), and the analysis produces pairs of chromatin interacting regions visualized in a heat map interaction frequency matrix. (2) Droplet Hi-C uses three different enzymes (DpnII, MboII, and NlaIII) to cut the DNA and, after ligation, uses a microfluidic system to obtain a cell-specific chromatin complex. (3) MicroC uses a double crosslinking step with FA and Disuccinimidyl Glutarate (DSG), represented by brown dots. Instead of a restriction enzyme, it uses an MNase to cut to a single-nucleosome resolution (~150pb). (4) HiCO 3D uses FA and an MNase. With modification to the bioinformatic analysis, the nucleosome orientation is defined, and the different colors in its matrix represent different nucleosome orientations. (5) C Walks uses FA and four cutters (such as DpnII) to obtain high-molecular-weight DNA that represents the concatenation of 30–60 DNA fragments that are subjected to shotgun sequencing. After computational assembly and analysis, the C-walks are defined

NuTi Capture-C for dispersed DNA elements as viewpoints [14], LI Capture-C and Tiled-C for low-cell inputs [15, 16], Tri-C for multi-way interactions [17] and Promoter-CHiC for genome-wide capture of gene promoter interactions [18–20].

Additionally, combining oligonucleotides with MNase has led to the development of MCC (MicroCapture-C), which uses an oligonucleotide-based enrichment approach to generate 3C interaction profiles at base-pair resolution [21]. Also, Tiled-MCC is an extension of MCC, which uses a dense tiling of capture oligonucleotides across large regions of interest, an approach that allows the generation of regional contact matrices at 20 bp and can resolve local nano-scale chromatin structures [22].

MC-4C (multi-contact 4C) is a new technique designed to study higher-order chromatin contact patterns at specific genomic sites. Like C-walks, this method uses large concatemers of proximity-ligated fragments for long-molecule sequencing, revealing multi-way chromatin interactions [23]. All these techniques have been crucial for revealing and understanding genomic topology. Essentially, they are variations of the same technique with improvements at different steps of the process (fixation, digestion, ligation, or analysis), and biases associated with these main steps exist in all of them [24, 25].

Notably, some of the 3C-based technologies have been taken to the single-cell level [29, 30]. Recently, an improvement of these single-cell approaches has been developed, called Droplet Hi-C, allowing the evaluation of single-cell chromatin interactions at better resolution using a microfluidics system [31].

2.2 Proximity Ligation or Fixation-Free Technologies

Other strategies that avoid proximity ligation and thus most of the biases presented in the 3C-based technologies have been developed and represent interesting alternatives.

GAM (Genome Architecture Mapping) is a technique that infers proximity matrices by combining nuclear cryosectioning with laser microdissection and DNA sequencing. In GAM, it is assumed that DNA regions in the same cryosection belong to the same interaction cluster [26] (Fig. 2a). SPRITE (split-pool recognition of interactions by tag extension) is a technique that uses a crosslinked chromatin and Split-and-pool strategy to isolate chromatin complexes. Adding sequential barcoding during the Split-and-pool procedure allows tagging of chromatin structures in the same complex with the same serial barcode [27] (Fig. 2b). ChIA-Drop (Chromatin Interaction Analysis using droplet-based and barcode-linked sequencing) uses cross-linked and fragmented chromatin with a microfluidic device; each chromatin complex is partitioned in an emulsion (GEM) for amplification and barcoding. After pooling for high-throughput sequencing and bioinformatic analysis, multiple chromatin interactions can be inferred [28] (Fig. 2c).

The methods above are ligation-free but fixation dependent. CAP-C uses a multifunctional chemical platform and UV irradiation instead of conventional fixation agents to eliminate crosslinking and digestion biases and generate high-resolution contact maps with low

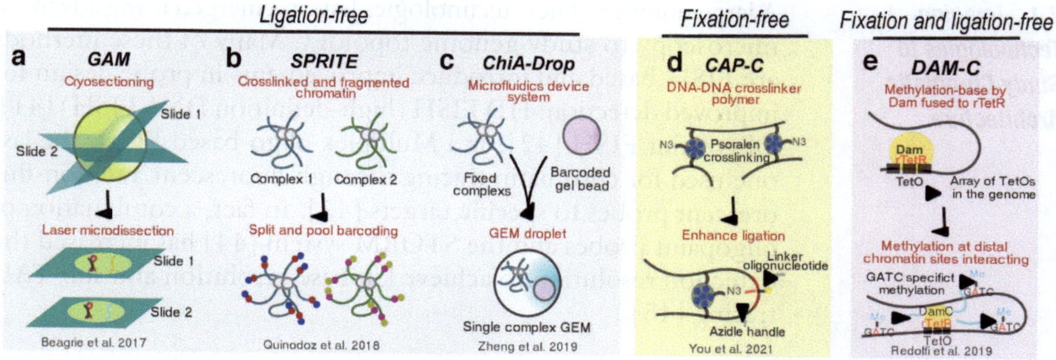

Fig. 2 Ligation or fixation-free 3D genome conformation technologies. (**a**) In GAM, fixed and frozen cells (green) are cryosectioned in thin slides (Slides 1 and 2) by laser microdissection. The DNA in each slide (DNA in the middle of the slides) is extracted, amplified and sequenced, and loci that are closer to each other in the nuclear space are detected in the same slide more often than distant loci. (**b**) In SPRITE, fixed and fragmented chromatin (complexes 1 and 2) is barcoded using a split-pool strategy in which interacting molecules within an individual complex are identified by sequencing and grouping reads that contain identical barcodes (color circles in each complex). (**c**) ChiA-Drop uses a Chromium microfluidics system (10× Genomics) in which a fixed and fragmented chromatin sample is directly loaded onto the microfluidics device, and each chromatin complex is partitioned in a gel-bead-in-emulsion (GEM) droplet that contains unique DNA oligonucleotides and reagents for linear amplification and barcoding (gel bed purple). (**d**) CAP-C uses spherical polymers (gray circles) with tens of crosslinking groups on the surface with diameters ranging from 3 nm to 11 nm to capture interactions in both closed and open chromatin conformations. Also, the polymer has an azide handle (N3) to react with a bifunctional bridge linker oligonucleotide (red) for proper two-step ligation. (**e**) DAM-C uses a Dam/rTeR fusion proteins (yellow circle) to methylate specific regions (GATC) in the genome (gray squares at DNA) through the insertion of an array of TetOs at the specific genomic viewpoint, inducing the recruitment of Dam molecules and methylation (Me)

background noise [32] (Fig. 2d); however, it is ligation-dependent. In this regard, DamC (DNA adenine methyltransferase C) represents the first crosslinking and ligation-free method to study the genome topology in vivo [33] (Fig. 2e). DamC is a modified version of DamID [34] in which a specific protein is fused to Dam methylase, and GATC sequences close to the fusion protein are methylated, enabling their identification by DNA sequencing.

2.3 Detection of DNA–RNA Interactions

In addition to DNA–DNA sequence-based methods, a lot of alternatives to detect RNA–DNA contacts genome-wide have emerged. For example, RD-SPRITE (RNA & DNA SPRITE) [35], MARGI (mapping RNA-genome interactions) [36], ChAR-seq (Chromatin-Associated RNA sequencing) [37], GRID-seq (Global RNA Interactions with DNA by deep sequencing) [38] and RADICL-seq (RNA And DNA Interacting Complexes Ligated and sequenced) [39], among others [40]. These technologies have allowed RNA annotation at different topological genomic structures.

2.4 Imaging Technologies to Study Chromatin Architecture

A large number of new technologies have reemerged using advanced microscopy to study genome topology. Many of these methods are FISH-based and introduce improvements in probe design for improved detection: HD FISH (high-definition DNA FISH) [41], Oligopaint FISH [42], and Multiplex oligo-based FISH, the last one used for chromatin tracing through fluorescent and non-fluorescent probes to specific targets [43]. In fact, a combination of oligopaint probes and the STORM system [44] has increased the detection resolution to achieve kilobase-resolution and sub-TAD tracing [45].

2.5 New Knowledge Arising from New Technologies

An interesting question in the chromatin architecture field is what the differences are between TADs and Loops. Finer-scale structures, such as stripes and dots in the interaction matrix, have been revealed thanks to the improvements in 3C-based technologies, such as Micro-C in mouse stem cells. These novel structures are caused by combinatorial binding of transcription factors, cofactors, architectural proteins and chromatin modifiers [46]. All these factors loop promoters-to-promoters (P–P) or enhancers-to-promoters (E–P) and bundle contacts between Polycomb regions inside of TADs, whereas CTCF and cohesin preferentially mediate long-range contacts (e.g., TADs). Interestingly, Hsieh et al. [48] described that upon acute loss of CTCF, cohesin, WAPL, or YY1 in mESC, there are minimal effects on short-range interactions after 3 h, even YY1, which is involved in E–P interactions [47]. They observed that using single-molecule imaging, cohesin depletion reduced transcription factors (TF) binding to chromatin, indicating that although CTCF, cohesin, WAPL, or YY1 are not important to short-term maintenance of E–P interactions, cohesin is involved in the TF binding to their targets more efficiently [48].

Also, Aljahani et al. used Tiled-MCC to study the effect of CTCF and cohesin depletion on nano-scale chromatin interactions in four specific regions [22]. They found that the loop extrusion mechanism is not essential for enhancer-promoter interactions, as these interactions were only slightly affected by CTCF and cohesin depletion [22].

In summary, it appears that TADs are primarily organized by structural proteins at their boundaries, while loops are organized by short interactions involving specific elements (e.g., transcription factors, coactivators, polymerase II) with promoters, enhancers, and super-enhancers, as well as the transcriptional state of the chromatin.

Additionally, using GAM, Beagrie et al. discovered an abundance of genome-wide three-way contacts between highly transcribed or super-enhancing regions, revealing a new layer of chromatin organization [26]. Using RD-SPRITE, Quinodoz et al. identified active and inactive interchromosomal hubs around nuclear speckles and nucleoli, demonstrating new higher-order interchromosomal interactions within the nucleus [27]. Furthermore, using CAP-C, You et al. detected transcription-state-dependent subcompartments that bring multiple domains into proximity, a phenomenon conserved from Drosophila to humans [32].

3 Genome 3D Organization in ESCs: Who Maintains the Pluripotent Interactome?

The genome topology of embryonic stem cells (ESCs) has been extensively studied to understand how chromatin structure is maintained and its importance in the pluripotent cell state and early development. Chromatin organization plays a key role in regulating ESCs; for instance, chromatin in ESCs is generally less condensed and more permissive, with low levels of DNA methylation and repressive chromatin marks. However, there are H3K27me3-repressed regions and bivalent domains at lineage-specification genes [49]. Additionally, decreased expression of cohesin in ESCs leads to premature differentiation, like the loss of the pluripotency factor OCT4 [50], highlighting the fundamental role of chromatin organization in maintaining ESC pluripotency. Also, new elements involved in maintaining the pluripotent state related to genome topology have been discovered.

3.1 Polycomb Complexes

Polycomb repressive complexes are key transcriptional regulators of developmental genes, and specifically, PRC1 and PRC2 maintain embryonic stem cell (ESC) pluripotency by silencing lineage-specifying developmental regulator genes and acting through the modulation of genome architecture [51, 52].

Using data from promoter capture Hi-C, 3C and DNA FISH of mouse ESC, Schoenfelder et al. found three-dimensional promoter–promoter and promoter–enhancer contact networks organized by PRC1 (mainly between poised enhancers), four Hox gene clusters and key developmental genes. Upon deletion of PRC1, genes in this network resulted in transcriptional upregulation and caused key cell fate decisions associated with early development [52].

Also, using 5C, stochastic optical reconstruction microscopy (STORM) and an "oligo-paint" strategy, Kundu et al. found that Polycomb Clusters create self-interacting domains (20–140 kb) that differ from TADs in ESCs and change upon differentiation to neural progenitor cells (NPCs) [53]. These repressive domains are disrupted upon removal of the Polyhomeotic subunit of the canonical PRC1, and interestingly, architectural proteins, including CTCF, are not found at the boundaries of PRC1 domains [53]. However, Boyle et al., using Ring1b-deletion in mESC, observed that there is a loss of interactions irrespective of whether the associated gene was up-regulated or not, suggesting that gene activation upon PRC1 subunit-deletion was not responsible for the loss of chromatin contacts in the absence of RING1B, and similarly that loss of interactions is not sufficient to activate genes [54].

On the other hand, one important question about the Polycom complex is: "*What is the mechanism of Polycomb recruitment to its target genes to form domains?.*"

Recently, Cuadrado et al. showed in mESC that cohesin variant SA2 facilitates Polycomb domain compaction through PRC1 recruitment and promotes the establishment of long-range interaction networks between distant Polycomb-bound promoters, and, in contrast, Cohesin-SA1 disrupts these networks while preserving TADs [55]. Also, Skourti-Stathaki et al. show that R-loops [56] are formed at a subset of Polycomb target genes (genes with well-annotated CpG islands in their promoters) and these R-loop-positive genes (e.g., Gata4), upon R-loop removal, lead to decreased PRC1 and PRC2 recruitment, and gene de-repression. In contrast, stable removal of PRC2 does not affect transcriptional repression of R-loop-positive genes, suggesting that Polycomb repression does not occur via one mechanism but consists of different layers of repression [57].

3.2 Pluripotency Factors

In ESC, the pluripotent state is mainly regulated by a group of transcription factors, including Sox2, Nanog, and Oct4, which are key to maintaining the pluripotent state [58].

Using 4C-seq in ESCs, it was previously reported that specific interactions between pluripotent factors and enrichment of Polycomb/H3K27me3 proteins occupy distinct nuclear spaces [59], suggesting an important role for cell-type-specific gene-regulatory networks in ESC genome organization [59]. Specifically, Nanog is one of the most studied transcription factors in ESC. Using a modified version of 4C-seq, Apostolou et al. found that the interaction network around the Nanog promoter was reorganized during differentiation and restored in induced pluripotent stem cells (iPSCs) [60]. This restoration led to a positive regulation of transcription in associated genes, which suggests a causal link.

Using promoter capture HiC (examining >22,000 promoters at single restriction fragment resolution) in mESC, Novo et al. identified interaction networks between super-enhancers (SEs) and the core of pluripotent factors, defining a specific chromatin structure organization of pluripotency [61] where Nanog is mainly responsible for these interactions. However, the deletion of Nanog only affects 20% of interactions, indicating that Nanog contributes to the stability of a subset of specific SE contacts in ESCs, particularly those interactions that span large distances [61]. Also, PHC1, a subunit of PCRC1, exerts its role in maintaining pluripotency through PRC1-independent Nanog activation by stabilizing genome-wide chromatin interactions at the Nanog locus [62], revealing further elements involved in the maintenance of pluripotency through specific chromatin interactions of pluripotent factors.

Another important transcription factor is Oct4, essential for maintaining pluripotency in vitro and in vivo [63, 64]. And, Sox2 (another pluripotency factor) has been reported with Oct4 to cooperatively bind to super-enhancers [61, 65, 66] and using fluorescent protein fusions of Oct4 and Sox2, DNA-dependent Oct4-Sox2 interactions at super-enhancers were discovered [66, 67]. Furthermore, Oct4 has been reported to be involved in chromatin accessibility in ESCs [68]. Oct4 binding events in pluripotent stem cells are required to keep chromatin accessible at their target sites [67], and this function is through Brg1, a chromatin remodeling complex [68]. OCT4 and BRG1 occupancy were highly correlated in the mouse ESC genome, and using ATAC-seq and BRG1-depleted cells, it was found that BRG1 is required to create accessible chromatin at OCT4 target sites in ESCs [68].

3.3 Architectural Proteins

As mentioned above, the chromatin structure in ESCs is generally less condensed and permissive than in lineage-committed cells, and this ESC genome topology changes during the differentiation [69]. In vertebrate genomes, stable chromatin loops are formed between CTCF that block Cohesin in a chromatin extrusion model [70].

CTCF and Cohesin are mammals' most important structural proteins [71]. For example, Nora et al. found that CTCF-depletion disappearance of TADs, uncovering CTCF as essential for maintaining TAD formation [72]. Smc1(subunit of cohesin) ChIA-PET showed that enhancers and genes maintaining the pluripotent identity of ESCs are insulated within loops anchored by both CTCF and cohesin [73].

On the other hand, recently, Su et al. identified a CTCF-binding element closest to the 3′- end of HoxA within the same TAD in ESC, and using CRISPR-Cas9-mediated deletion of the binding site element upregulated HoxA expression and enhanced early ESC differentiation [74]. Capture-C was done, revealing that deletion of the binding site decreased interactions between adjacent enhancers and central regions of HoxA chromatin. Also, has been reported that CTCF associates with the critical pluripotent NANOG-DPPA3-GDF3 locus, and CTCF-depletion accelerates the loss of Pluripotency in hESC by expression decrease of pluripotency genes, and this correlates with decreased CTCF binding in a particular site where CTCF and cofactors act as enhancer barriers [75]. Interestingly, Bridge Linker-Hi-C [76] experiments in mESC showed that CTCF maintains chromatin accessibility and mediates enhancer-promoter loop formation for pluripotency genes during reprogramming to induce the pluripotent state [77].

Also, Rad21 (subunit of cohesin) ChIP-seq analysis in ESCs has revealed a subset of cohesin binding sites, co-occupied by pluripotency transcription factors and the removal of Rad21 causes similar defects as Nanog-deficient cells, indicating cohesin has an important role in establishing/maintaining pluripotency [78].

Wapl is a protein that dissociates cohesin rings from chromatin [79], and this is important for controlling cohesin levels on chromosomes [80]. Using auxin-inducible degron (AID) of endogenous Wapl protein in mESC, it showed that acute depletion of WAPL (or Cohesin) leads to differentiation by cohesin redistribution or cohesin loss, and regions of depleted cohesin are enriched for Sox2, Oct4, Nanog, and enhancers leading to downregulation of pluripotency factors [81]. All these results highlight the important role of structural proteins in the genome organization of ESC.

4 Genome Organization During Lineage Commitment from ESCs

During lineage commitment, genome organization is crucial for many specific key molecular events in each kind of cell type, including the spatiotemporal regulation of the transcription of specific cell type genes and regulatory elements [82].

For example, Dixon et al. observed that during the differentiation of human ESCs, TADs seem to be conserved between ESCs and the differentiated tissues; however, numerous changes in chromatin structure occur within domains (intra-TADs) and genes within domains that have increased intra-domain interaction frequency tend to be upregulated, while genes within domains that decrease intra-domain interaction frequency tend to be downregulated [83]. In this sense, specific interactions (mediating by specific elements) have been discovered as important for lineage commitment, for example, in SE (surface epithelium) commitment, GRHL3 (grainy head-like 3) was identified as a key initiation factor sufficient to drive SE commitment mediating interactions [84], also during T lineage commitment, there are changes of intra-TAD interactions and the increase in intra-TAD connectivity preceded the expression onset of the enclosing genes [85] and during the erythroid differentiation using Tiled-C [16] Oudelaar et al. identified specific interactions between regulatory elements that progressively increase during differentiation [86]. Also, Fraser et al., through neuronal differentiation and using HiC, concluded that the topologies of "metaTAD trees", which represent large-scale contacts between nearby TADs, undergo a degree (~20%) of structural reorganization during the differentiation of mouse ESCs to neural progenitor cells [87].

In this way, although the TADs seem to be conserved between ESCs and the differentiated tissues, the intra-TAD and inter-TAD interactions are very dynamic and key to lineage commitment.

On the other hand, it has been observed that the compartments change during differentiation. For example, in human ESCs differentiation, 36% of the genome switches to another compartment in at least one of the lineages analyzed [83]. Also, comparing ESCs and differentiated endothelial cells (ECs), around ~2000 compartments underwent a switch from inactive to active or active to inactive [88].

5 Mis-Regulation of Genome Organization During Development

As discussed so far, the dynamic organization of the genome is essential to regulate transcription programs, maintenance, and development of pluripotent cells. This correct organization must be maintained to avoid alterations in gene expression and development, given that disruption of this organization can lead to different diseases such as congenital malformation syndromes [89]. Alterations in TAD structure can conduce to the novo long-range interactions between promoters and enhancers, producing aberrant gene expression patterns and severe phenotypes in development [89, 90], indeed, structural variations (SVs) as deletions, inversions, duplications, and translocations have been associated with gene misexpression and diseases through TADs disruption, leading to specific ectopic aberrant enhancer-promoter interactions [90]. For example, *Pitx1* is a gene that is critically required to establish the identity and differentiation of hindlimbs in vertebrates [91], and is regulated by an enhancer (*Pen*) that displays activity in both forelimbs and hindlimbs, but, the enhancer only interacts with Pitx1 in the hindlimb, leading to upregulation, enabling the *Pen* enhancer to control *Pitx1* transcription [92]. However, if the forelimb-specific 3D chromatin conformation is disrupted and allows the interaction between *Pen* and *Pitx1*, *Pitx1* is disrupted, and its misexpression in forelimbs leads to limb malformation diseases [93].

5.1 Structural Variations at the Epha4 Gene Locus

Different congenital diseases have been reported as a consequence of TAD disruption, for example, one of the most cited works about this is Lupiañez et al. where they achieved resemble brachydactyly, F-syndrome/syndactyly, and polydactyly in mouse using genetic engineering (CRISPR/Cas9) to induce inversions, duplications, and deletions, modifying TADs structures in the *EPHA4* locus [94], highlighting the importance of TAD boundaries in gene regulation as an interaction regulatory unit that encompass enhancers and promoters, and also highlighting the importance of the SVs to modified these boundaries elements and leads to development-diseases.

5.2 Hematopoietic Disorders

More recently, the role of the CTCF boundaries has been reinforced by Owens et al., who used Tiled-C and observed that the deletions of CTCF sites in a sub-TAD led to a partial loss of domain structure, mildly affecting gene expression of *Runx1* [95] (Fig. 3). *Runx1* plays a critical role in the de novo generation of the hematopoietic system and maintenance of normal hematopoietic homeostasis [95]. The CTCF site deletion using CRISPR/Cas9 leads to reduced sub-TAD boundary strengths and delays hematopoietic development [95], highlighting the importance of CTCF boundaries in the maintenance of chromatin structure and gene expression of hematopoietic homeostasis.

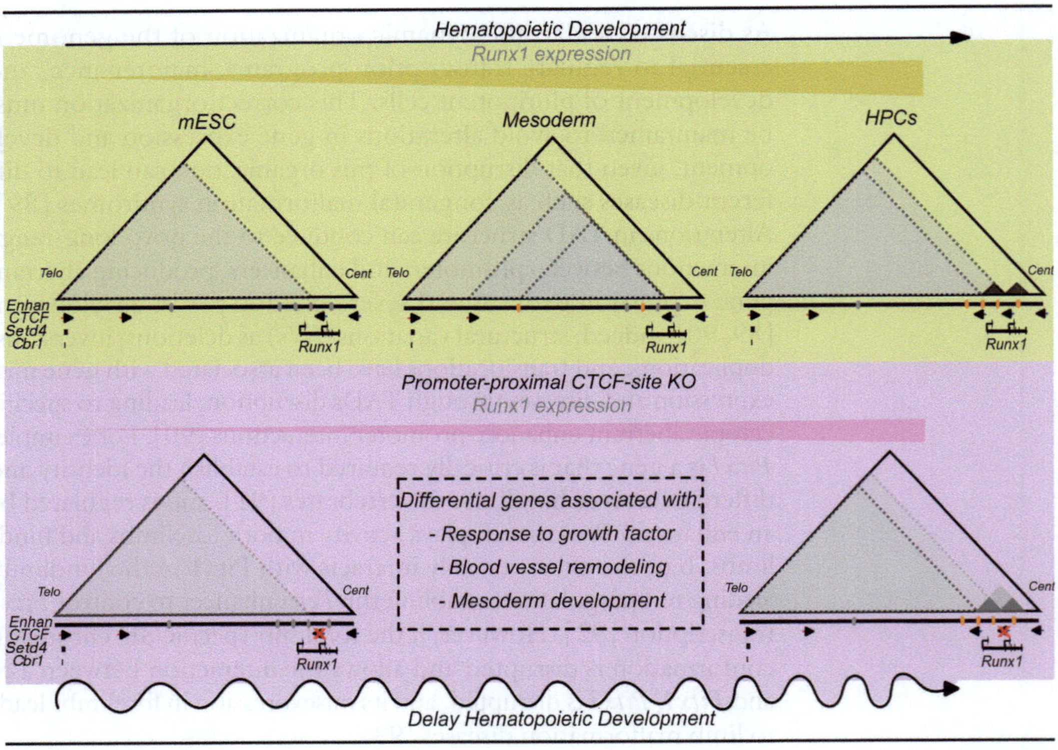

Fig. 3 Runx1 topology during hematopoietic differentiation from mESC. Large light gray triangles represent the 1.1 Mb *Runx1* TAD, and throughout the TAD, dashed lines represent long-range chromatin interactions, with the darker color indicating stronger interactions inside the TAD. At the top, the TAD structure during normal differentiation from mESC (left) to mesoderm (middle) and HPCs (hematopoietic progenitor cells on the right) is shown. The orientation of selected CTCF motifs is indicated in red and blue arrows below the TADs, also active (orange circle) and inactive (gray circle) enhancers, and the genes Setd4, Cbr1 and Runx1 are shown. The telomeric side of the TAD is shown on the left, and the centromeric side is on the right. Runx1 gene expression increases during differentiation, as indicated. On the bottom, a specific CTCF-site close to the Runx1-promoter was deleted (red X on CTCF-site) in mESC (CTCF-KO mESC), and these cells were differentiated to HPCs (CTCF-KO HPCs), leading to Runx1 expression downregulation compared to WT HPCs. In the middle, Gene Ontologies of the differentially expressed genes between the CTCF-KO and the WT HPCs are shown. In HPCs, the Runx1 sub-TADs emerge during differentiation, but in CTCF-KO HPCs, intra-sub-TADs interactions decrease while the inter-sub-TADs interactions increase. The curved line at the bottom represents the delay of the hematopoietic development in the CTCF-KO condition

5.3 Split-Hand/Foot Malformation Type 3

There are new papers that continue to reinforce the importance of the TAD structures in developmental diseases, for example, Cova et al., using Capture HiC, studied a congenital malformation termed SHFM3 (Split-Hand/Foot Malformation type 3), and they focus on an *LBX1/FGF8* locus that is considered the cause for SHFM type 3 [96].

This locus is organized in a major chromatin domain encompassing two distinct smaller chromatin regulatory domains, where the centromeric and the telomeric domains contain the *Lbx1/Btrc* and the *Fbxw4/Fgf8* genes. These two domains are separated by CTCF boundaries, and the *Fgf8* TAD contains several well-characterized tissue-specific enhancers essential to the *Fgf8* expression pattern in the embryo (Fig. 4a). *LBX1* is an essential gene for early muscle cell differentiation and migration [97], and *FGF8* is expressed together with other FGFs in the AER (apical ectodermal ridge) where it functions as a growth factor for the underlying mesenchyme [98].

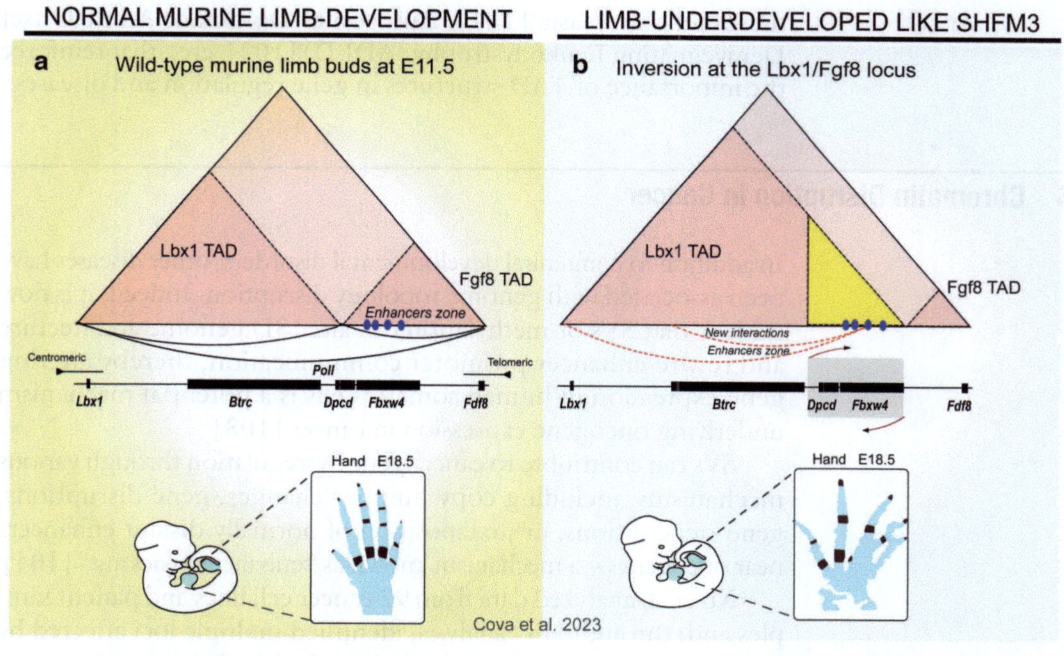

Fig. 4 Topological structure alteration in limb development. **a**) Wild-type topology in murine limb bud cells at E11.5 days. *Lbx1* and *Fgf8* are located within distinct TADs (two TADs inside of a large TAD) separated by boundaries. The Fgf8 TAD contains several enhancers in an enhancer zone (blue circles), and the Lbx1 TAD contains the Btrc gene (Black boxes) indicated below the TADs. The centromeric and telomeric side is indicated with black arrows. Below the TADs, the interactions between TAD-boundaries (black lines) are indicated. At the bottom, skeletal analysis of fore- and hindlimb stained with alcian blue (cartilage) and alizarin red (bone) from wild-type at E18.5 is shown. Zoom of the normal development hand is shown on the right (inside of the blank square). **b**) SHFM3-phenotype is revealed through CRISPR/Cas9 inversion induction (boxes, shadow, and red arrows). The inversion spans (yellow TAD-region) the Fgf8-enhancers zone and a short region of Lbx1-TAD, leading to relocalization of enhancers and a larger Lbx1-TAD with new interactions between enhancers and Lbx1 and Btrc (dashed red lines). At the bottom, the skeletal malformation is shown at E18.5 day as a result of the inversion, resembling the SHFM3 syndrome

There are reported duplications in SHFM3 that encompass five genes of the locus (*LBX1*, *BTRC*, *POLL*, *DPCD*, and *FBXW4*), excluding the neighboring *FGF8* [99]. Also, Cova et al. identified an inversion that could be associated with this disease, by which they engineered the duplication and inversion at the *Lbx1/Fgf8* locus in mice. Using the CRISPR/Cas9 genome editing tool and Capture HiC, they found that only the homozygous inversion results in TAD-restructuration, where ectopic 3D interactions led to the SHFM3 phenotype [96] (Fig. 4b). The new interactions lead to ectopic activation of the *Lbx1* and *Btrc* genes, along with downregulation of the Fgf8 in the AER, mediated by specific enhancer relocalization. Interestingly, the isolated disruption of each gene does not lead to the SHFM3 phenotype, but rather, the combinatorial disruption of these genes produces the SHFM3 phenotype, highlighting the importance of the different complex mechanisms by which TAD disruption can induce the combinatorial disruption of genes and contribute to specific diseases.

Further literature has been published about TAD disruption, leading to different diseases such as Liebenberg syndrome [100], mesomelic dysplasia [101], Autosomal-Dominant Adult-Onset Demyelinating Leukodystrophy (ADLD) [102], etc., that reinforce the importance of TAD structures in gene regulation and diseases.

6 Chromatin Disruption in Cancer

In addition to congenital developmental disorders, other diseases have been associated with genome topology disruption. Indeed, it is now evident that SVs or methylation can alter 3D genome architecture and rewire enhancer-promoter communication, thereby altering gene expression in human somatic cells is a potential mechanism underlying oncogene expression in cancer [103].

SVs can contriubte to cancer gene dysregulation through various mechanisms, including copy number changes, gene disruptions, gene–gene fusions, or juxtaposition of normally distant enhancers near oncogenes—a mechanism known as "enhancer hijacking" [104].

Xu et al. analyzed data from 92 cancer cell lines and patient samples and, through HiC analysis, identified multiple loci affected by recurrent regulatory 3D genome alterations in diverse cancer types. Among these, some rearrangements included TADs associated with poorer prognosis [103]. They also found 80.7% of SVs result in TAD fusions, signaling that most rearrangements create breakpoints that cross TAD boundaries, and aberrantly connect oncogenes such as *MYC*, *TERT*, and *CCND1* with regulatory elements [103].

6.1 TAD Fusion

Gastro-Intestinal Stromal Tumors (GISTs) are the most common gastrointestinal tract sarcoma. A subset of GISTs have succinate dehydrogenase (SDH) deficiency and global DNA hyper-methylation [105].

Flavahan et al. found hundreds of putative CTCF DNA binding sites where DNA methylation prevents CTCF binding and using Hi-C, they showed that this lack of CTCF binding affected two TADs, the first is a 250 kb TAD that contains *FGF3* and *FGF4* and the second adjacent TAD contains a large cluster of enhancers. They showed that the hyper-methylation affects the binding sites of CTCF at the TAD boundary that partitions this super-enhancer from the FGF genes, and also mediated 4C analysis; they demonstrated that the super-enhancer physically interacted beyond ~200 kb of the boundary, contacting with the *FGF3* and *FGF4* oncogenes [106] (Fig. 5).

Then, using CRISPR-Cas9, they deleted the CTCF sites between the two TADs, resulting in a six-fold induction of FGF4 expression and a 35-fold induction of FGF3 expression. In conclusion, in SDH-deficient GIST, the methylation or deletion of CTCF sites leads to the TAD boundary misfunction and crosstalk interactions between TADs, inducing TADs fusion and expression of oncogenes like FGF3 and FGF4 [106, 107].

6.2 Enhancer Hijacking

Using a machine learning-based method, Kim et al. investigated SV-mediated gene dysregulation by profiling 3D cancer genome maps from 40 patients with colorectal cancer (CRC). Building Hi-C contact maps, they identified a total of 562 *cis*- and 235 *trans*-de novo chromatin contacts as a consequence of SVs in CRCs, and interestingly, the genes associated with these chromatin contacts tended to be upregulated. They identified high-confidence interactions between promoters and enhancers, mapping to 1695 genes targeted by relocated super enhancers (SEs) [108].

It is widely accepted that enhancer-hijacking is mediated by SV-linked TAD fusions or re-shuffling of regulatory elements between two adjacent TADs [109]. However, in this work, they found that their enhancer-hijacking events span multiple TADs, and this mechanism can generate overexpression of the proto-oncogene *CCDC117*, which is associated with DNA damage repair and cell-cycle progression [110]. Indeed, those events span multiple TADs associated with weak TAD boundaries, and often the SVs that cause them are also found in the active compartment [108] (Fig. 5).

6.3 Silencer Hijacking

Enhancer hijacking caused by topological defects has been characterized in various cancer types as a promoter of malignancy [111–113]. Recently, silencer hijacking has also been reported [114].

Acute myeloid leukemia (AML) is a myeloid neoplasm characterized by differentiation blockade and clonal proliferation. A recent publication analyzed 25 AML samples and 7 healthy donor samples and performed HiC and 4C [4, 5, 7]; they found 200 AML-specific chromatin loops with 220 genes that were significantly upregulated and 88 genes that were downregulated. These loops contain enhancers and silencers, of which 42.4% of AML-specific loops were between

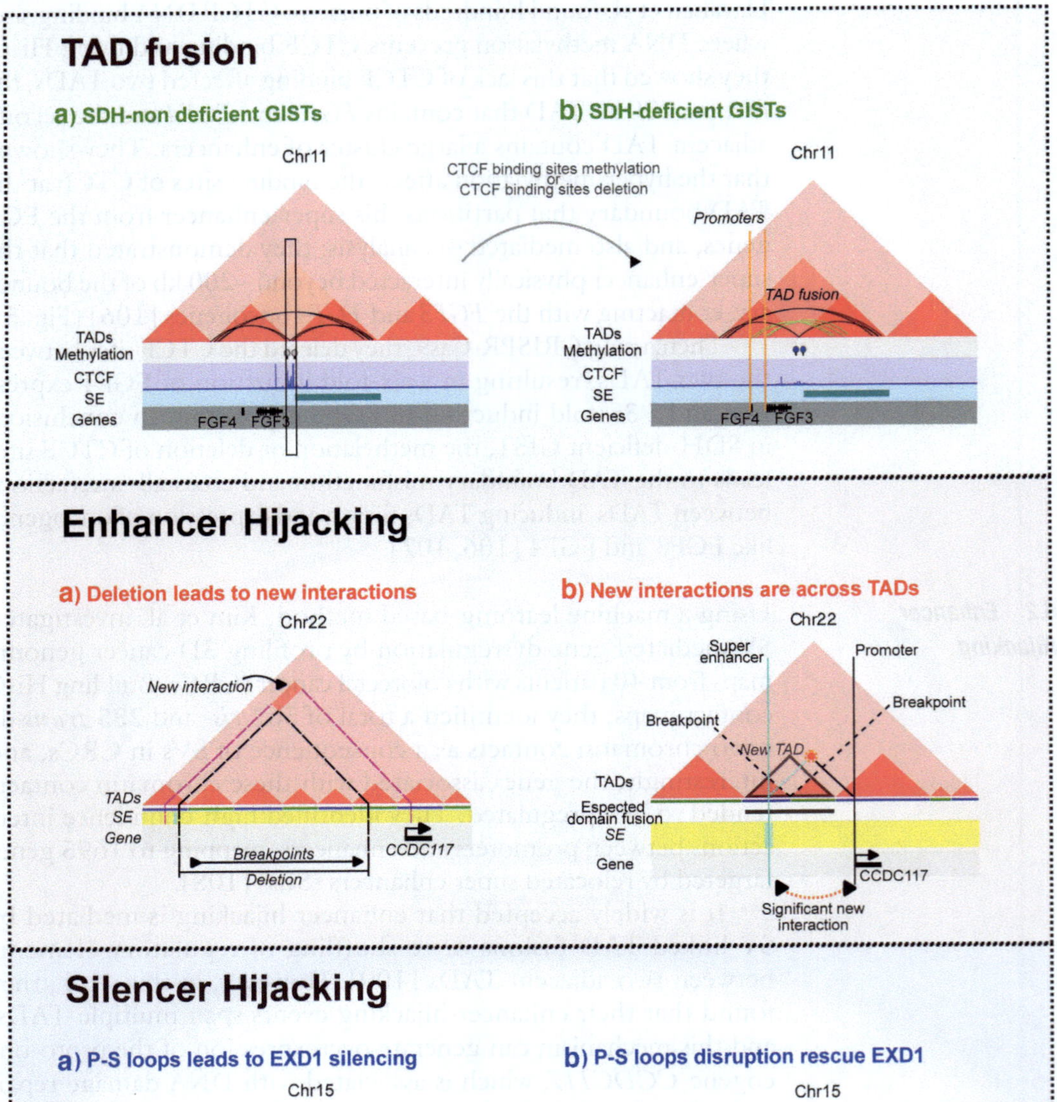

Fig. 5 TADs-disruption in cancer. TAD fusion mechanism. **a)** Chromatin structure in chromosome 11 is shown in SDH-non-deficient GISTs. The TADs are shown in the Hi-C matrix (red triangles), and specific interactions between them (black lines). Below the TADs, two CpG sites in a non-methylated state (gray circles), CTCF-binding sites at the specific sites (blue signal), and the super-enhancer (SE) (blue bar) in the TAD next to the FGF3-TAD (on the right) are shown. On the bottom, the genes are shown in a specific TAD (black boxes for

gene promoters and enhancers (P–E) and 11.2% were between promoters and silencers (P–S) (Fig. 5). Also, in 25 AML samples, they used Hi-C and WSG to call neo-loops formed by SVs. Overall, 5.7% of all the neo-loops led to silencer hijacking, and 17.2% of the neo-loops led to enhancer hijacking. Concomitantly, genes with hijacked enhancers had a significantly higher expression, whereas genes with hijacked silencers showed decreased expression. Examples of down-regulated genes included AML-related genes (*JAK1* and *KMT2C*) [114]. Also, to validate the hijacked silencers, in Kasumi-1 cells they used the CRISPR/dCas9–UTX (histone demethylase) system to remove chromatin methylation of a group of silencers (de-repression) that were predicted to regulate *EXD1* (*EXD*1 silencing), and after treatment, the contacts between the silencer region and *EXD1* disappeared and *EXD1* expression increased, validating the repressive role of these hijacked silencers on their target genes and uncovering a novel silencing mechanism in cancer [114] (Fig. 5).

7 Concluding Remarks

To this day, many exciting technologies to characterize genome 3D organization have been developed and continue to evolve, increasing and detailing our knowledge of the mechanisms behind genome architecture and the consequences of their failure in disease. Improvements in the basic steps during 3C-based technologies, as well as novel strategies that do not rely on fixation or proximity-ligation, provide new signals to measure and quantify chromosome conformations in cell populations and in single cells.

Fig. 5 (continued) *FGF3* and *FGF4*). **b**) Chromatin structure in SDH-deficient GISTs. SDH-deficient GISTs show methylation at the two CpG sites in *FGF3* boundary (blue circles), in consequence, less CTCF binding at this boundary is shown, leading to inter-TADs interactions (green lines) that involve *FGF3* and *FGF4* promoters (orange vertical lines) and SE. Enhancer Hijacking. **a**) Chromatin structure at Chromosome 22 (Chr22) in CRC patients. The specific deletion (black dashed lines) that spans several TADs (red triangles) at Chr22 is shown in the Hi-C matrix, leading to new interactions (pink dashed lines) above TADs. Specific new interactions between SE and the CCDC117 gene promoter are shown (red rectangle). **b**) The restructured Hi-C matrix resulting from the deletion is shown. Two specific TADs are fused at their breakpoints (black dashed lines), leading to significant specific new interactions occurring between a SE (blue vertical line) and the CCDC117 gene promoter (black vertical lines). Silencer Hijacking. **a**) Chromatin structure at Chr15 in Acute myeloid leukemia (AML) patients. **a**) An inversion at Chr15 (green and brown arrows) leads to EXD1 downregulation (Black boxes correspond to EXD1 gene position). Specific EXD1-promoter interactions with the neighboring TAD are shown. H3K27me3-enrichment (black circles between breakpoints with high red signal) is termed Silencer-Promoter interactions. **b**) Using the CRISPR dCas9–UTX system (gray molecule that binds DNA with a yellow UTX), the H3H27me3 enrichment is removed, leading to reduced signal of H3K27me3 (black block-line) and EXD1 gene upregulation

Genome architecture plays a pivotal role in cell potency and identity determination, from compartments, micro-compartments, TADs, and chromatin loops connecting enhancers and silencers to gene promoters. Defects in the molecular machinery needed to build these topological layers lead to developmental abnormalities and cell transformation. Consequently, genome editing and CRISPR/dCas9 tools, together with the technologies to measure genome architecture at high resolution, have become essential to mimic these chromatin disruptions and to characterize the diverse topological disruption mechanisms and their functional consequences in disease.

Acknowledgments

This work was supported by CONAHCyT (now SECIHTI) grants 303068, 15758, 137721, and PAPPIT grant IN210323. E.B.O. was supported by SECIHTI scholarship 777256.

Conflict of Interest The authors have no conflicts of interest to declare that are relevant to the content of this chapter.

References

1. Lamond AI, Earnshaw WC (1998) Structure and function in the nucleus. Science 280(5363):547–553
2. Cremer T, Cremer C (2001) Chromosome territories, nuclear architecture and gene regulation in mammalian cells. Nat Rev Genet 2(4):292–301
3. Dekker J, Rippe K, Dekker M, Kleckner N (2002) Capturing chromosome conformation. Science 295(5558):1306–1311
4. Simonis M, Klous P, Splinter E, Moshkin Y, Willemsen R, de Wit E, van Steensel B, de Laat W (2006) Nuclear organization of active and inactive chromatin domains uncovered by chromosome conformation capture-on-chip (4C). Nat Genet 38(11):1348–1354
5. Zhao Z, Tavoosidana G, Sjölinder M, Göndör A, Mariano P, Wang S, Kanduri C, Lezcano M, Sandhu KS, Singh U, Pant V, Tiwari V, Kurukuti S, Ohlsson R (2006) Circular chromosome conformation capture (4C) uncovers extensive networks of epigenetically regulated intra- and interchromosomal interactions. Nat Genet 38(11):1341–1347
6. Dostie J, Richmond TA, Arnaout RA, Selzer RR, Lee WL, Honan TA, Rubio ED, Krumm A, Lamb J, Nusbaum C, Green RD, Dekker J (2006) Chromosome Conformation Capture Carbon Copy (5C): a massively parallel solution for mapping interactions between genomic elements. Genome Res 16(10):1299–1309
7. Lieberman-Aiden E, van Berkum NL, Williams L, Imakaev M, Ragoczy T, Telling A, Amit I, Lajoie BR, Sabo PJ, Dorschner MO, Sandstrom R, Bernstein B, Bender MA, Groudine M, Gnirke A, Stamatoyannopoulos J, Mirny LA, Lander ES, Dekker J (2009) Comprehensive mapping of long range interactions reveals folding principles of the human genome. Science 326(5950):289–293
8. Hsieh TH, Weiner A, Lajoie B, Dekker J, Friedman N, Rando OJ (2015) Mapping nucleosome resolution chromosome folding in yeast by Micro-C. Cell 162(1):108–119
9. Krietenstein N, Abraham S, Venev SV, Abdennur N, Gibcus J, Hsieh TS, Parsi KM, Yang L, Maehr R, Mirny LA, Dekker J, Rando OJ (2020) Ultrastructural details of mammalian chromosome architecture. Mol Cell 78(3):554–565
10. Akgol Oksuz B, Yang L, Abraham S, Venev SV, Krietenstein N, Parsi KM, Ozadam H, Oomen ME, Nand A, Mao H, Genga RMJ, Maehr R, Rando OJ, Mirny LA, Gibcus JH, Dekker J (2021) Systematic evaluation of chromosome conformation capture assays. Nat Methods 18(9):1046–1055

11. Ohno M, Ando T, Priest DG, Taniguchi Y (2021) Hi-CO: 3D genome structure analysis with nucleosome resolution. Nat Protoc 16(7):3439–3469
12. Olivares-Chauvet P, Mukamel Z, Lifshitz A, Schwartzman O, Elkayam NO, Lubling Y, Deikus G, Sebra RP, Tanay A (2016) Capturing pairwise and multi-way chromosomal conformations using chromosomal walks. Nature 540(7632):296–300
13. Downes DJ, Smith AL, Karpinska MA, Velychko T, Rue-Albrecht K, Sims D, Milne TA, Davies JOJ, Oudelaar AM, Hughes JR (2022) Capture-C: a modular and flexible approach for high-resolution chromosome conformation capture. Nat Protoc 17(2):445–475
14. Downes DJ, Beagrie RA, Gosden ME, Telenius J, Carpenter SJ, Nussbaum L, De Ornellas S, Sergeant M, Eijsbouts CQ, Schwessinger R, Kerry J, Roberts N, Shivalingam A, El-Sagheer A, Oudelaar AM, Brown T, Buckle VJ, Davies JOJ, Hughes JR (2021) High-resolution targeted 3C interrogation of cis-regulatory element organization at genome-wide scale. Nat Commun 12(1):531
15. Oudelaar AM, Davies JOJ, Downes DJ, Higgs DR, Hughes JR (2017) Robust detection of chromosomal interactions from small numbers of cells using low-input Capture-C. Nucleic Acids Res 45(22):e184–e184
16. Oudelaar AM, Beagrie RA, Gosden M, de Ornellas S, Georgiades E, Kerry J, Hidalgo D, Carrelha J, Shivalingam A, El-Sagheer AH, Telenius JM, Brown T, Buckle VJ, Socolovsky M, Higgs DR, Hughes JR (2020) Dynamics of the 4D genome during in vivo lineage specification and differentiation. Nat Commun 11(1):2722
17. Oudelaar AM, Davies JOJ, Hanssen LLP, Telenius JM, Schwessinger R, Liu Y, Brown JM, Downes DJ, Chiariello AM, Bianco S, Nicodemi M, Buckle VJ, Dekker J, Higgs DR, Hughes JR (2018) Single-allele chromatin interactions identify regulatory hubs in dynamic compartmentalized domains. Nat Genet 50(12):1744–1751
18. Schoenfelder S, Furlan-Magaril M, Mifsud B, Tavares-Cadete F, Sugar R, Javierre BM, Nagano T, Katsman Y, Sakthidevi M, Wingett SW, Dimitrova E, Dimond A, Edelman LB, Elderkin S, Tabbada K, Darbo E, Andrews S, Herman B, Higgs A, LeProust E, Osborne CS, Mitchell JA, Luscombe NM, Fraser P (2015) The pluripotent regulatory circuitry connecting promoters to their long-range interacting elements. Genome Res 25(4):582–597
19. Mifsud B, Tavares-Cadete F, Young AN, Sugar R, Schoenfelder S, Ferreira L, Wingett SW, Andrews S, Grey W, Ewels PA, Herman B, Happe S, Higgs A, LeProust E, Follows GA, Fraser P, Luscombe NM, Osborne CS (2015) Mapping long-range promoter contacts in human cells with high-resolution capture Hi-C. Nat Genet 47(6):598–606
20. Schoenfelder S, Javierre BM, Furlan-Magaril M, Wingett SW, Fraser P (2018) Promoter capture Hi-C: high-resolution, genome-wide profiling of promoter interactions. J Vis Exp 136:57320
21. Hamley JC, Li H, Denny N, Downes D, Davies JOJ (2023) Determining chromatin architecture with Micro Capture-C. Nat Protoc 18(6):1687–1711
22. Aljahani A, Hua P, Karpinska MA, Quililan K, Davies JOJ, Oudelaar AM (2022) Analysis of sub-kilobase chromatin topology reveals nano-scale regulatory interactions with variable dependence on cohesin and CTCF. Nat Commun 13(1):2139
23. Vermeulen C, Allahyar A, Bouwman BAM, Krijger PHL, Verstegen MJAM, Geeven G, Valdes-Quezada C, Renkens I, Straver R, Kloosterman WP, de Ridder J, de Laat W (2020) Multi-contact 4C: long-molecule sequencing of complex proximity ligation products to uncover local cooperative and competitive chromatin topologies. Nat Protoc 15(2):364–397
24. Yaffe E, Tanay A (2011) Probabilistic modeling of Hi-C contact maps eliminates systematic biases to characterize global chromosomal architecture. Nat Genet 43(11):1059–1065
25. Chandradoss KR, Guthikonda PK, Kethavath S, Dass M, Singh H, Nayak R, Kurukuti S, Sandhu KS (2020) Biased visibility in Hi-C datasets marks dynamically regulated condensed and decondensed chromatin states genome-wide. BMC Genomics 21(1):175
26. Beagrie RA, Scialdone A, Schueler M, Kraemer DC, Chotalia M, Xie SQ, Barbieri M, de Santiago I, Lavitas LM, Branco MR, Fraser J, Dostie J, Game L, Dillon N, Edwards PA, Nicodemi M, Pombo A (2017) Complex multi-enhancer contacts captured by genome architecture mapping. Nature 543(7646):519–524
27. Quinodoz SA, Ollikainen N, Tabak B, Palla A, Schmidt JM, Detmar E, Lai MM, Shishkin AA, Bhat P, Takei Y, Trinh V, Aznauryan E, Russell P, Cheng C, Jovanovic M, Chow A, Cai L, McDonel P, Garber M, Guttman M (2018) Higher-order inter-chromosomal hubs shape 3D genome organization in the nucleus. Cell 174(3):744–757
28. Zheng M, Tian SZ, Capurso D, Kim M, Maurya R, Lee B, Piecuch E, Gong L, Zhu JJ, Li Z, Wong CH, Ngan CY, Wang P, Ruan X, Wei CL, Ruan Y (2019) Multiplex chromatin

interactions with single-molecule precision. Nature 566(7745):558–562
29. Nagano T, Lubling Y, Stevens TJ, Schoenfelder S, Yaffe E, Dean W, Laue ED, Tanay A, Fraser P (2013) Single-cell Hi-C reveals cell-to-cell variability in chromosome structure. Nature 502(7469):59–64
30. Nagano T, Lubling Y, Várnai C, Dudley C, Leung W, Baran Y, Mendelson Cohen N, Wingett S, Fraser P, Tanay A (2017) Cell-cycle dynamics of chromosomal organization at single-cell resolution. Nature 547(7661):61–67
31. Chang L, Xie Y, Taylor B, Wang Z, Sun J, Armand EJ, Mishra S, Xu J, Tastemel M, Lie A, Gibbs ZA, Indralingam HS, Tan TM, Bejar R, Chen CC, Furnari FB, Hu M, Ren B (2025) Droplet Hi-C enables scalable, single-cell profiling of chromatin architecture in heterogeneous tissues. Nat Biotechnol 43(10):1694–1707
32. You Q, Cheng AY, Gu X, Harada BT, Yu M, Wu T, Ren B, Ouyang Z, He C (2021) Direct DNA crosslinking with CAP-C uncovers transcription-dependent chromatin organization at high resolution. Nat Biotechnol 39(2):225–235
33. Redolfi J, Zhan Y, Valdes-Quezada C, Kryzhanovska M, Guerreiro I, Iesmantavicius V, Pollex T, Grand RS, Mulugeta E, Kind J, Tiana G, Smallwood SA, de Laat W, Giorgetti L (2019) DamC reveals principles of chromatin folding in vivo without crosslinking and ligation. Nat Struct Mol Biol 26(6):471–480
34. van Steensel B, Henikoff S (2000) Identification of in vivo DNA targets of chromatin proteins using tethered dam methyltransferase. Nat Biotechnol 18(4):424–428
35. Quinodoz SA, Jachowicz JW, Bhat P, Ollikainen N, Banerjee AK, Goronzy IN, Blanco MR, Chovanec P, Chow A, Markaki Y, Thai J, Plath K, Guttman M (2021) RNA promotes the formation of spatial compartments in the nucleus. Cell 184(23):5775–5790
36. Sridhar B, Rivas-Astroza M, Nguyen TC, Chen W, Yan Z, Cao X, Hebert L, Zhong S (2017) Systematic mapping of RNA-chromatin interactions in vivo. Curr Biol 27(4):602–609
37. Bell JC, Jukam D, Teran NA, Risca VI, Smith OK, Johnson WL, Skotheim JM, Greenleaf WJ, Straight AF (2018) Chromatin-associated RNA sequencing (ChAR-seq) maps genome-wide RNA-to-DNA contacts. Elife 7:e27024
38. Li X, Zhou B, Chen L, Gou LT, Li H, Fu XD (2017) GRID-seq reveals the global RNA–chromatin interactome. Nat Biotechnol 35(10):940–950
39. Bonetti A, Agostini F, Suzuki AM, Hashimoto K, Pascarella G, Gimenez J, Roos L, Nash AJ, Ghilotti M, Cameron CJF, Valentine M, Medvedeva YA, Noguchi S, Agirre E, Kashi K, Samudyata, Luginbühl J, Cazzoli R, Agrawal S, Luscombe NM, Blanchette M, Kasukawa T, de Hoon M, Arner E, Lenhard B, Plessy C, Castelo-Branco G, Orlando V, Carninci P (2020) RADICL-seq identifies general and cell type-specific principles of genome-wide RNA–chromatin interactions. Nat Commun 11(1):1018
40. Xiao Q, Huang X, Zhang Y, Xu W, Yang Y, Zhang Q, Hu Z, Xing F, Sun Q, Li G, Li X (2022) The landscape of promoter-centred RNA–DNA interactions in rice. Nat Plants 8(2):157–170
41. Bienko M, Crosetto N, Teytelman L, Klemm S, Itzkovitz S, van Oudenaarden A (2013) A versatile genome-scale PCR-based pipeline for high-definition DNA FISH. Nat Methods 10(2):122–124
42. Beliveau BJ, Joyce EF, Apostolopoulos N, Yilmaz F, Fonseka CY, McCole RB, Chang Y, Li JB, Senaratne TN, Williams BR, Rouillard JM, Wu CT (2012) Versatile design and synthesis platform for visualizing genomes with Oligopaint FISH probes. Proc Natl Acad Sci USA 109(52):21301–21306
43. Hu M, Wang S (2021) Chromatin tracing: imaging 3D genome and nucleome. Trends Cell Biol 31(1):5–8
44. Rust MJ, Bates M, Zhuang X (2006) Sub-diffraction-limit imaging by stochastic optical reconstruction microscopy (STORM). Nat Methods 3(10):793–795
45. Bintu B, Mateo LJ, Su JH, Sinnott-Armstrong NA, Parker M, Kinrot S, Yamaya K, Boettiger AN, Zhuang X (2018) Super-resolution chromatin tracing reveals domains and cooperative interactions in single cells. Science 362(6413):eaau1783
46. Hsieh TS, Cattoglio C, Slobodyanyuk E, Hansen AS, Rando OJ, Tjian R, Darzacq X (2020) Resolving the 3D landscape of transcription-linked mammalian chromatin folding. Mol Cell 78(3):539–553. e8
47. Weintraub AS, Li CH, Zamudio AV, Sigova AA, Hannett NM, Day DS, Abraham BJ, Cohen MA, Nabet B, Buckley DL, Guo YE, Hnisz D, Jaenisch R, Bradner JE, Gray NS, Young RA (2017) YY1 is a structural regulator of enhancer-promoter loops. Cell 171(7):1573–1588
48. Hsieh TS, Cattoglio C, Slobodyanyuk E, Hansen AS, Darzacq X, Tjian R (2022) Enhancer–promoter interactions and transcription are largely maintained upon acute loss of CTCF, cohesin, WAPL or YY1. Nat Genet 54(12):1919–1932
49. Pelham-Webb B, Murphy D, Apostolou E (2020) Dynamic 3D chromatin reorganization during establishment and maintenance of pluripotency. Stem Cell Rep 15(6):1176–1195

50. Kagey MH, Newman JJ, Bilodeau S, Zhan Y, Orlando DA, van Berkum NL, Ebmeier CC, Goossens J, Rahl PB, Levine SS, Taatjes DJ, Dekker J, Young RA (2010) Mediator and cohesin connect gene expression and chromatin architecture. Nature 467(7314):430–435
51. Simon JA, Kingston RE (2013) Occupying chromatin: Polycomb mechanisms for getting to genomic targets, stopping transcriptional traffic, and staying put. Mol Cell 49(5):808–824
52. Schoenfelder S, Sugar R, Dimond A, Javierre BM, Armstrong H, Mifsud B, Dimitrova E, Matheson L, Tavares-Cadete F, Furlan-Magaril M, Segonds-Pichon A, Jurkowski W, Wingett SW, Tabbada K, Andrews S, Herman B, LeProust E, Osborne CS, Koseki H, Fraser P, Luscombe NM, Elderkin S (2015) Polycomb repressive complex PRC1 spatially constrains the mouse embryonic stem cell genome. Nat Genet 47(10):1179–1186
53. Kundu S, Ji F, Sunwoo H, Jain G, Lee JT, Sadreyev RI, Dekker J, Kingston RE (2017) Polycomb repressive complex 1 generates discrete compacted domains that change during differentiation. Mol Cell 65(3):432–446. e5
54. Boyle S, Flyamer IM, Williamson I, Sengupta D, Bickmore WA, Illingworth RS (2020) A central role for canonical PRC1 in shaping the 3D nuclear landscape. Genes Dev 34(13–14):931–949
55. Cuadrado A, Giménez-Llorente D, Kojic A, Rodríguez-Corsino M, Cuartero Y, Martín-Serrano G, Gómez-López G, Marti-Renom MA, Losada A (2019) Specific contributions of cohesin-SA1 and cohesin-SA2 to TADs and polycomb domains in embryonic stem cells. Cell Rep 27(12):3500–3510. e4
56. Aguilera A, García-Muse T (2012) R loops: from transcription byproducts to threats to genome stability. Mol Cell 46(2):115–124
57. Skourti-Stathaki K, Torlai Triglia E, Warburton M, Voigt P, Bird A, Pombo A (2019) R-loops enhance polycomb repression at a subset of developmental regulator genes. Mol Cell 73(5):930–945
58. Yeo JC, Ng HH (2013) The transcriptional regulation of pluripotency. Cell Res 23(1):20–32
59. Denholtz M, Bonora G, Chronis C, Splinter E, de Laat W, Ernst J, Pellegrini M, Plath K (2013) Long-range chromatin contacts in embryonic stem cells reveal a role for pluripotency factors and polycomb proteins in genome organization. Cell Stem Cell 13(5):602–616
60. Apostolou E, Ferrari F, Walsh RM, Bar-Nur O, Stadtfeld M, Cheloufi S, Stuart HT, Polo JM, Ohsumi TK, Borowsky ML, Kharchenko PV, Park PJ, Hochedlinger K (2013) Genome-wide chromatin interactions of the Nanog locus in pluripotency, differentiation, and reprogramming. Cell Stem Cell 12(6):699–712
61. Novo CL, Javierre BM, Cairns J, Segonds-Pichon A, Wingett SW, Freire-Pritchett P, Furlan-Magaril M, Schoenfelder S, Fraser P, Rugg-Gunn PJ (2018) Long-range enhancer interactions are prevalent in mouse embryonic stem cells and are reorganized upon pluripotent state transition. Cell Rep 22(10):2615–2627
62. Chen L, Tong Q, Chen X, Jiang P, Yu H, Zhao Q, Sun L, Liu C, Gu B, Zheng Y, Fei L, Jiang X, Li W, Volpe G, Abdul MM, Guo G, Zhang J, Qian P, Sun Q, Neculai D, Esteban MA, Li C, Wen F, Ji J (2021) PHC1 maintains pluripotency by organizing genome-wide chromatin interactions of the Nanog locus. Nat Commun 12(1):2829
63. Niwa H, Miyazaki J, Smith AG (2000) Quantitative expression of Oct-3/4 defines differentiation, dedifferentiation or self-renewal of ES cells. Nat Genet 24(4):372–376
64. Nichols J, Zevnik B, Anastassiadis K, Niwa H, Klewe-Nebenius D, Chambers I, Schöler H, Smith A (1998) Formation of pluripotent stem cells in the mammalian embryo depends on the POU transcription factor Oct4. Cell 95(3):379–391
65. Whyte WA, Orlando DA, Hnisz D, Abraham BJ, Lin CY, Kagey MH, Rahl PB, Lee TI, Young RA (2013) Master transcription factors and mediator establish super-enhancers at key cell identity genes. Cell 153(2):307–319
66. Lam CS, Mistri TK, Foo YH, Sudhaharan T, Gan HT, Rodda D, Lim LH, Chou C, Robson P, Wohland T, Ahmed S (2012) DNA-dependent Oct4–Sox2 interaction and diffusion properties characteristic of the pluripotent cell state revealed by fluorescence spectroscopy. Biochem J 448(1):21–33
67. Xiong L, Tolen EA, Choi J, Velychko S, Caizzi L, Velychko T, Adachi K, MacCarthy CM, Lidschreiber M, Cramer P, Schöler HR (2022) Oct4 differentially regulates chromatin opening and enhancer transcription in pluripotent stem cells. Elife 11:e71533
68. King HW, Klose RJ (2017) The pioneer factor OCT4 requires the chromatin remodeller BRG1 to support gene regulatory element function in mouse embryonic stem cells. Elife 6:e22631
69. Barrington C, Georgopoulou D, Pezic D, Varsally W, Herrero J, Hadjur S (2019) Enhancer accessibility and CTCF occupancy underlie asymmetric TAD architecture and cell type specific genome topology. Nat Commun 10(1):2908

70. Rowley MJ, Corces VG (2018) Organizational principles of 3D genome architecture. Nat Rev Genet 19(12):789–800
71. Sanborn AL, Rao SS, Huang SC, Durand NC, Huntley MH, Jewett AI, Bochkov ID, Chinnappan D, Cutkosky A, Li J, Geeting KP, Gnirke A, Melnikov A, McKenna D, Stamenova EK, Lander ES, Aiden EL (2015) Chromatin extrusion explains key features of loop and domain formation in wild-type and engineered genomes. Proc Natl Acad Sci USA 112(47):E6456–E6465
72. Nora EP, Goloborodko A, Valton AL, Gibcus JH, Uebersohn A, Abdennur N, Dekker J, Mirny LA, Bruneau BG (2017) Targeted degradation of CTCF decouples local insulation of chromosome domains from genomic compartmentalization. Cell 169(5):930–944. e22
73. Dowen JM, Fan ZP, Hnisz D, Ren G, Abraham BJ, Zhang LN, Weintraub AS, Schujiers J, Lee TI, Zhao K, Young RA (2014) Control of cell identity genes occurs in insulated neighborhoods in mammalian chromosomes. Cell 159(2):374–387
74. Su G, Wang W, Chen J, Liu M, Zheng J, Guo D, Bi J, Zhao Z, Shi J, Zhang L, Lu W (2021) CTCF-binding element regulates ESC differentiation via orchestrating long-range chromatin interaction between enhancers and HoxA. J Biol Chem 296:100413
75. Balakrishnan SK, Witcher M, Berggren TW, Emerson BM (2012) Functional and molecular characterization of the role of CTCF in human embryonic stem cell biology. PLoS One 7(8):e42424
76. Liang Z, Li G, Wang Z, Djekidel MN, Li Y, Qian MP, Zhang MQ, Chen Y (2017) BL-Hi-C is an efficient and sensitive approach for capturing structural and regulatory chromatin interactions. Nat Commun 8(1):1622
77. Song Y, Liang Z, Zhang J, Hu G, Wang J, Li Y, Guo R, Dong X, Babarinde IA, Ping W, Sheng YL, Li H, Chen Z, Gao M, Chen Y, Shan G, Zhang MQ, Hutchins AP, Fu XD, Yao H (2022) CTCF functions as an insulator for somatic genes and a chromatin remodeler for pluripotency genes during reprogramming. Cell Rep 39(1):110626
78. Nitzsche A, Paszkowski-Rogacz M, Matarese F, Janssen-Megens EM, Hubner NC, Schulz H, de Vries I, Ding L, Huebner N, Mann M, Stunnenberg HG, Buchholz F (2011) RAD21 cooperates with pluripotency transcription factors in the maintenance of embryonic stem cell identity. PLoS One 6(5):e19470
79. Huis in 't Veld PJ, Herzog F, Ladurner R, Davidson IF, Piric S, Kreidl E, Bhaskara V, Aebersold R, Peters JM (2014) Characterization of a DNA exit gate in the human cohesin ring. Science 346(6212):968–972
80. Kueng S, Hegemann B, Peters BH, Lipp JJ, Schleiffer A, Mechtler K, Peters JM (2006) Wapl controls the dynamic association of cohesin with chromatin. Cell 127(5):955–967
81. Liu NQ, Maresca M, van den Brand T, Braccioli L, Schijns MMGA, Teunissen H, Bruneau BG, Nora EP, de Wit E (2021) WAPL maintains a cohesin loading cycle to preserve cell-type-specific distal gene regulation. Nat Genet 53(1):100–109
82. Zheng H, Xie W (2019) The role of 3D genome organization in development and cell differentiation. Nat Rev Mol Cell Biol 20(9):535–550
83. Dixon JR, Jung I, Selvaraj S, Shen Y, Antosiewicz-Bourget JE, Lee AY, Ye Z, Kim A, Rajagopal N, Xie W, Diao Y, Liang J, Zhao H, Lobanenkov VV, Ecker JR, Thomson JA, Ren B (2015) Chromatin architecture reorganization during stem cell differentiation. Nature 518(7539):331–336
84. Huang H, Liu J, Li M, Guo H, Zhu J, Zhu L, Wu S, Mo K, Huang Y, Tan J, Chen C, Wang B, Yu Y, Wang L, Liu Y, Ouyang H (2022) Cis-regulatory chromatin loops analysis identifies GRHL3 as a master regulator of surface epithelium commitment. Sci Adv 8(28):eabo5668
85. Hu G, Cui K, Fang D, Hirose S, Wang X, Wangsa D, Jin W, Ried T, Liu P, Zhu J, Rothenberg EV, Zhao K (2018) Transformation of accessible chromatin and 3D nucleome underlies lineage commitment of early T cells. Immunity 48(2):227–242. e8
86. Oudelaar AM, Beagrie RA, Gosden M, de Ornellas S, Georgiades E, Kerry J, Hidalgo D, Carrelha J, Shivalingam A, El-Sagheer AH, Telenius JM, Brown T, Buckle VJ, Socolovsky M, Higgs DR, Hughes JR (2020) Dynamics of the 4D genome during in vivo lineage specification and differentiation. Nat Commun 11(1):2722
87. Fraser J, Ferrai C, Chiariello AM, Schueler M, Rito T, Laudanno G, Barbieri M, Moore BL, Kraemer DC, Aitken S, Xie SQ, Morris KJ, Itoh M, Kawaji H, Jaeger I, Hayashizaki Y, Carninci P, Forrest AR, FANTOM Consortium, Semple CA, Dostie J, Pombo A, Nicodemi M (2015) Hierarchical folding and reorganization of chromosomes are linked to transcriptional changes in cellular differentiation. Mol Syst Biol 11(12):852
88. Niskanen H, Tuszynska I, Zaborowski R, Heinäniemi M, Ylä-Herttuala S, Wilczynski B, Kaikkonen MU (2018) Endothelial cell differentiation is encompassed by changes in long

range interactions between inactive chromatin regions. Nucleic Acids Res 46(4):1724–1740
89. Anania C, Lupiáñez DG (2020) Order and disorder: abnormal 3D chromatin organization in human disease. Brief Funct Genomics 19(2):128–138
90. Sivakumar A, de Las Heras JI, Schirmer EC (2019) Spatial genome organization: from development to disease. Front Cell Dev Biol 7:18
91. DeLaurier A, Schweitzer R, Logan M (2006) Pitx1 determines the morphology of muscle, tendon, and bones of the hindlimb. Dev Biol 299(1):22–34
92. Kragesteen BK, Spielmann M, Paliou C, Heinrich V, Schöpflin R, Esposito A, Annunziatella C, Bianco S, Chiariello AM, Jerković I, Harabula I, Guckelberger P, Pechstein M, Wittler L, Chan WL, Franke M, Lupiáñez DG, Kraft K, Timmermann B, Vingron M, Visel A, Nicodemi M, Mundlos S, Andrey G (2018) Dynamic 3D chromatin architecture contributes to enhancer specificity and limb morphogenesis. Nat Genet 50(10):1463–1473
93. Spielmann M, Brancati F, Krawitz PM, Robinson PN, Ibrahim DM, Franke M, Hecht J, Lohan S, Dathe K, Nardone AM, Ferrari P, Landi A, Wittler L, Timmermann B, Chan D, Mennen U, Klopocki E, Mundlos S (2012) Homeotic arm-to-leg transformation associated with genomic rearrangements at the PITX1 locus. Am J Hum Genet 91(4):629–635
94. Lupiáñez DG, Kraft K, Heinrich V, Krawitz P, Brancati F, Klopocki E, Horn D, Kayserili H, Opitz JM, Laxova R, Santos-Simarro F, Gilbert-Dussardier B, Wittler L, Borschiwer M, Haas SA, Osterwalder M, Franke M, Timmermann B, Hecht J, Spielmann M, Visel A, Mundlos S (2015) Disruptions of topological chromatin domains cause pathogenic rewiring of gene-enhancer interactions. Cell 161(5):1012–1025
95. Owens DDG, Anselmi G, Oudelaar AM, Downes DJ, Cavallo A, Harman JR, Schwessinger R, Bucakci A, Greder L, de Ornellas S, Jeziorska D, Telenius J, Hughes JR, de Bruijn MFTR (2022) Dynamic Runx1 chromatin boundaries affect gene expression in hematopoietic development. Nat Commun 13(1):773
96. Cova G, Glaser J, Schöpflin R, Prada-Medina CA, Ali S, Franke M, Falcone R, Federer M, Ponzi E, Ficarella R, Novara F, Wittler L, Timmermann B, Gentile M, Zuffardi O, Spielmann M, Mundlos S (2023) Combinatorial effects on gene expression at the Lbx1/Fgf8 locus resolve split-hand/foot malformation type 3. Nat Commun 14(1):1475
97. Buckingham M (2001) Skeletal muscle formation in vertebrates. Curr Opin Genet Dev 11(4):440–448
98. Niswander L (2002) Interplay between the molecular signals that control vertebrate limb development. Int J Dev Biol 46(7):877–881
99. de Mollerat XJ, Gurrieri F, Morgan CT, Sangiorgi E, Everman DB, Gaspari P, Amiel J, Bamshad MJ, Lyle R, Blouin JL, Allanson JE, Le Marec B, Wilson M, Braverman NE, Radhakrishna U, Delozier-Blanchet C, Abbott A, Elghouzzi V, Antonarakis S, Stevenson RE, Munnich A, Neri G, Schwartz CE (2003) A genomic rearrangement resulting in a tandem duplication is associated with split hand–split foot malformation 3 (SHFM3) at 10q24. Hum Mol Genet 12(16):1959–1971
100. Spielmann M, Lupiáñez DG, Mundlos S (2018) Structural variation in the 3D genome. Nat Rev Genet 19(7):453–467
101. Flöttmann R, Wagner J, Kobus K, Curry CJ, Savarirayan R, Nishimura G, Yasui N, Spranger J, Van Esch H, Lyons MJ, DuPont BR, Dwivedi A, Klopocki E, Horn D, Mundlos S, Spielmann M (2015) Microdeletions on 6p22.3 are associated with mesomelic dysplasia Savarirayan type. J Med Genet 52(7):476–483
102. Giorgio E, Robyr D, Spielmann M, Ferrero E, Di Gregorio E, Imperiale D, Vaula G, Stamoulis G, Santoni F, Atzori C, Gasparini L, Ferrera D, Canale C, Guipponi M, Pennacchio LA, Antonarakis SE, Brussino A, Brusco A (2015) A large genomic deletion leads to enhancer adoption by the lamin B1 gene: a second path to autosomal dominant adult-onset demyelinating leukodystrophy (ADLD). Hum Mol Genet 24(11):3143–3154
103. Xu Z, Lee DS, Chandran S, Le VT, Bump R, Yasis J, Dallarda S, Marcotte S, Clock B, Haghani N, Cho CY, Akdemir KC, Tyndale S, Futreal PA, McVicker G, Wahl GM, Dixon JR (2022) Structural variants drive context-dependent oncogene activation in cancer. Nature 612(7940):564–572
104. Weischenfeldt J, Ibrahim DM (2023) When 3D genome changes cause disease: the impact of structural variations in congenital disease and cancer. Curr Opin Genet Dev 80:102048
105. Killian JK, Kim SY, Miettinen M, Smith C, Merino M, Tsokos M, Quezado M, Smith Jr WI, Jahromi MS, Xekouki P, Szarek E, Walker RL, Lasota J, Raffeld M, Klotzle B, Wang Z, Jones L, Zhu Y, Wang Y, Waterfall JJ, O'Sullivan MJ, Bibikova M, Pacak K, Stratakis C, Janeway KA, Schiffman JD, Fan JB, Helman L, Meltzer PS (2013) Succinate dehydrogenase mutation underlies global

epigenomic divergence in gastrointestinal stromal tumor. Cancer Discov 3(6):648–657
106. Flavahan WA, Drier Y, Johnstone SE, Hemming ML, Tarjan DR, Hegazi E, Shareef SJ, Javed NM, Raut CP, Eschle BK, Gokhale PC, Hornick JL, Sicinska ET, Demetri GD, Bernstein BE (2019) Altered chromosomal topology drives oncogenic programs in SDH-deficient GISTs. Nature 575(7781):229–233
107. Arao T, Ueshima K, Matsumoto K, Nagai T, Kimura H, Hagiwara S, Sakurai T, Haji S, Kanazawa A, Hidaka H, Iso Y, Kubota K, Shimada M, Utsunomiya T, Hirooka M, Hiasa Y, Toyoki Y, Hakamada K, Yasui K, Kumada T, Toyoda H, Sato S, Hisai H, Kuzuya T, Tsuchiya K, Izumi N, Arii S, Nishio K, Kudo M (2013) FGF3/FGF4 amplification and multiple lung metastases in responders to sorafenib in hepatocellular carcinoma. Hepatology 57(4):1407–1415
108. Kim K, Kim M, Lee AJ, Song SH, Kang JK, Eom J, Kang GH, Bae JM, Min S, Kim Y, Lim Y, Kim HS, Kim YJ, Kim TY, Jung I (2023) Spatial and clonality-resolved 3D cancer genome alterations reveal enhancer-hijacking as a potential prognostic marker for colorectal cancer. Cell Rep 42(7):112778
109. Valton AL, Dekker J (2016) TAD disruption as oncogenic driver. Curr Opin Genet Dev 36:34–40
110. Horton AJ, Brooker J, Streitfeld WS, Flessa ME, Pillai B, Simpson R, Clark CD, Gooz MB, Sutton KK, Foley AC, Lee KH (2019) Nkx2-5 Second Heart Field Target Gene Ccdc117 Regulates DNA Metabolism and Proliferation. Sci Rep 9(1):1738
111. Drier Y, Cotton MJ, Williamson KE, Gillespie SM, Ryan RJ, Kluk MJ, Carey CD, Rodig SJ, Sholl LM, Afrogheh AH, Faquin WC, Queimado L, Qi J, Wick MJ, El-Naggar AK, Bradner JE, Moskaluk CA, Aster JC, Knoechel B, Bernstein BE (2016) An oncogenic MYB feedback loop drives alternate cell fates in adenoid cystic carcinoma. Nat Genet 48(3):265–272
112. Montefiori LE, Bendig S, Gu Z, Chen X, Pölönen P, Ma X, Murison A, Zeng A, Garcia-Prat L, Dickerson K, Iacobucci I, Abdelhamed S, Hiltenbrand R, Mead PE, Mehr CM, Xu B, Cheng Z, Chang TC, Westover T, Ma J, Stengel A, Kimura S, Qu C, Valentine MB, Rashkovan M, Luger S, Litzow MR, Rowe JM, den Boer ML, Wang V, Yin J, Kornblau SM, Hunger SP, Loh ML, Pui CH, Yang W, Crews KR, Roberts KG, Yang JJ, Relling MV, Evans WE, Stock W, Paietta EM, Ferrando AA, Zhang J, Kern W, Haferlach T, Wu G, Dick JE, Klco JM, Haferlach C, Mullighan CG (2021) Enhancer hijacking drives oncogenic BCL11B expression in lineage-ambiguous stem cell leukemia. Cancer Discov 11(11):2846–2867
113. Zhang X, Choi PS, Francis JM, Imielinski M, Watanabe H, Cherniack AD, Meyerson M (2016) Identification of focally amplified lineage-specific super-enhancers in human epithelial cancers. Nat Genet 48(2):176–182
114. Xu J, Song F, Lyu H, Kobayashi M, Zhang B, Zhao Z, Hou Y, Wang X, Luan Y, Jia B, Stasiak L, Wong JH, Wang Q, Jin Q, Jin Q, Fu Y, Yang H, Hardison RC, Dovat S, Platanias LC, Diao Y, Yang Y, Yamada T, Viny AD, Levine RL, Claxton D, Broach JR, Zheng H, Yue F (2022) Subtype-specific 3D genome alteration in acute myeloid leukaemia. Nature 611(7935):387–398

Subject Index

Paul Delgado Olguin (ed.), *Mouse Embryogenesis*, Methods in Molecular Biology, vol. 3010,
https://doi.org/10.1007/978-1-0716-5126-1,

MIX
Papier aus verantwortungsvollen Quellen
Paper from responsible sources
FSC® C105338

If you have any concerns about our products,
you can contact us on
ProductSafety@springernature.com

In case Publisher is established outside the EU,
the EU authorized representative is:
Springer Nature Customer Service Center GmbH
Europaplatz 3, 69115 Heidelberg, Germany

Printed by Libri Plureos GmbH
in Hamburg, Germany